CONTENTS

CONTENTS

Think Academy Math Workbook is written with the hopes of sharing the beauty and excitement found in the study of mathematics. It is designed to not only teach students math knowledge, but also to train their logical thinking skills and abilities. Through logical thinking, hands-on projects and teacher support, Think Academy aims to enrich and enliven the learning of math.

What makes this book special?

1 Three Tracks

The course materials are differentiated by three levels based on difficulties; namely, Think Olympiad, Honors, and Advanced. The book enables teachers to deliver effective student-centered teaching methods based on students' existing skill levels.

Think Academy student learning goals	Advanced	Honors	Think Olympiad
Learning outcome	5 on all STEM-related AP subject tests; receive A's in all honor courses; GPA above 4.0 in Math; collegiate level thinking and problem-solving ability	Math becomes competitive advantage in college admission; qualify for AIME before Grade 9	Qualify for AIME by Grade 8 and receive a score of 10 on AIME; prepare to advance to USA(J)MO and IMO

2 *Four Modules*

Students are able to systematically learn math through four modules, and build up their own knowledge base as they progress through each one.

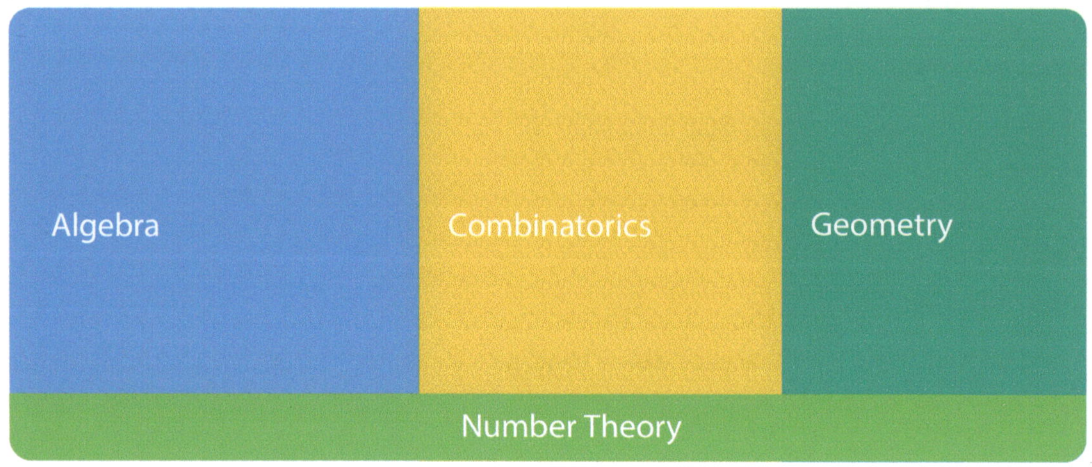

3 *Year-round Curriculum*

This book is created with a thorough understanding of how children from middle school learn. It provides specific solutions on how to systematically improve their math skills using the following curriculum structure.

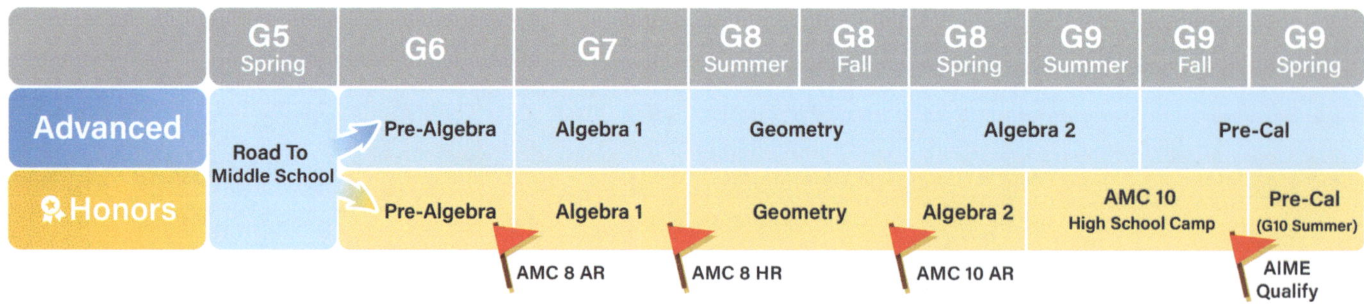

Lesson 1 Absolute-Value Equations and Inequalities

Learning Objectives

1. Simplify the absolute value by casework.

2. Solve absolute-value equations.

3. Solve absolute-value inequalities.

The **absolute value** can also be written as:

$$|a| = \begin{cases} a & a \geq 0 \\ -a & a < 0 \end{cases}$$

One way to simplify absolute values is by finding the zeros of the expression inside the absolute value sign. This method works because, at zeros of the expression, the expression changes signs.

Math Exploration 1

Answer the following questions:
(1) When $x \geq 1$, simplify $2|x - 1|$: _______ .

$$2x - 2$$

(2) When $x < 1$, simplify $2|x - 1|$: _______ .

$$-2x + 2$$

Answer the following questions:

(1) When $x \geqslant 1$, simplify $1 + 3|x - 1|$: __B__ .

    A. $3x - 3$ B. $3x - 2$ C. $3x + 3$ D. $3x + 4$

(2) When $x < 1$, simplify $1 + 3|x - 1|$: __D__ .

    A. $3x - 2$ B. $3x - 3$ C. $3 - 3x$ D. $4 - 3x$

For an absolute-value equation $|ax + b| = c$ (a, b and c are all real numbers), when $c \geqslant 0$, the original equation can be separated into two new equations $ax + b = c$ or $ax + b = -c$ (a, b and c are all real numbers);

when $c < 0$, there is no solution for this equation as the absolute value of any real number must be no less than 0.

The Value of c	The Solution of Absolute-Value Equations
$c > 0$	$ax + b = c$ or $ax + b = -c$
$c = 0$	$ax + b = 0$
$c < 0$	No solution

Math Exploration 2

Solve the following absolute-value equations.

(1) $|3x + 2| = -1$ no solution

$x = -1$

(2) $|2x - 4| = 0$

$x = 2$

(3) $\dfrac{|2x - 1|}{2} - 3 = 0$

$x = 3.5$

Solve the following absolute-value equations.

(1) $|x + 5| = 3$

    A. $x = -8$ B. $x = -2$

    C. $x = -8$ or $x = -2$ D. No solution

$= C$

(2) $2|x + 1| + 4 = 0$

    A. $x = 1$ B. $x = -3$

    C. $x = 1$ or $x = -3$ D. No solution

$= D$

For the inequality $|ax + b| < c$, it can be changed to $-c < ax + b < c$.

INEQUALITY	EQUIVALENT FORM	GRAPH
$|ax+b|<c$	$-c<ax+b<c$	
$|ax+b|\leq c$	$-c\leq ax+b\leq c$	

The general steps to solve an absolute-value inequality:

1. Rewrite the absolute value equation as compound inequalities that do not have absolute-value symbols.

2. Solve the compound inequalities as usual.

3. Write out the final solution.

Example:

Solve $|2x + 1| < 3$.

Step 1: Rewrite the absolute value equation as compound inequalities.

$\underline{-3} < 2x + 1 < \underline{3}$.

Step 2: Solve the compound inequalities.

$\underline{-2} < x < \underline{1}$.

Step 3: Write the final solution.

The solution is $\underline{-2} < x < \underline{1}$.

1 Draw the solution of the absolute-value inequality $|x| < 2$ on the number line.

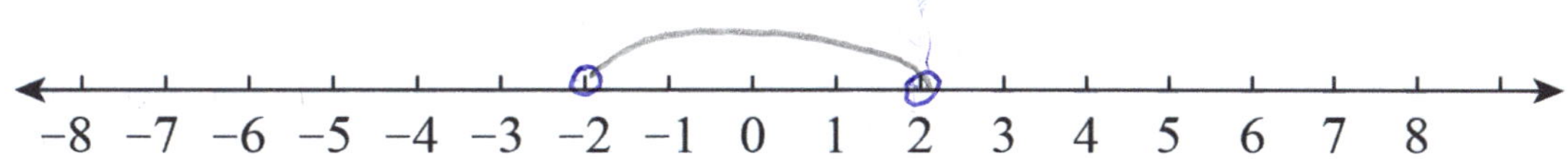

2 Solve the absolute-value inequalities $|x - 2| \leqslant 1$.

$$-1 \leqslant x - 2 \leqslant 1$$
$$1 \leqslant x \leqslant 3$$

3 Solve the absolute-value inequalities $\left|\dfrac{2x - 1}{3}\right| \leqslant 2$.

$$|2x - 1| \leqslant 6$$
$$-6 \leqslant 2x - 1 \leqslant 6$$
$$-5 \leqslant 2x \leqslant 7$$
$$-5/2 \leqslant x \leqslant 7/2$$

$$-\frac{5}{2} \leqslant x \leqslant \frac{7}{2}$$

1 Solve the absolute-value inequality $|5 - 2x| \leqslant 2$.

A. $\dfrac{3}{2} \leqslant x \leqslant \dfrac{7}{2}$ B. $x \leqslant \dfrac{3}{2}$ C. $x \geqslant \dfrac{7}{2}$ D. $x \leqslant \dfrac{3}{2}$ or $x \geqslant \dfrac{7}{2}$

$$-2 \leqslant 5 - 2x \leqslant 2$$
$$-7 \leqslant -2x \leqslant -3$$
$$-\frac{7}{2} \leqslant -x \leqslant -\frac{3}{2}$$
$$\frac{7}{2} \geqslant x \geqslant \frac{3}{2}$$

2 Solve the absolute-value inequality $\left|\dfrac{1}{2}x + 2\right| < 9$.

$$-9 < \frac{1}{2}x + 2 < 9$$
$$-11 < \frac{1}{2}x < 7$$
$$-22 < x < 14$$

3 Solve the absolute-value inequality: $-3|x - 1| + 3 > 0$.

$$-3|x - 1| > -3$$
$$3|x - 1| < 3$$
$$|x - 1| < 1$$
$$-1 < x - 1 < 1$$
$$0 < x < 2$$

◎ *Concept 4: Solving an Absolute-Value Inequality with ">"*

For the inequality $|ax + b| > c$, it can be changed to $ax + b < -c$ or $ax + b > c$.

INEQUALITY	EQUIVALENT FORM	GRAPH
$|ax+b|>c$	$ax+b<-c \ or \ ax+b>c$	
$|ax+b|\geq c$	$ax+b\leq -c \ or \ ax+b\geq c$	

Math Exploration 4

1 Draw the solution of the absolute-value inequality $|x| > 3$ on the number line.

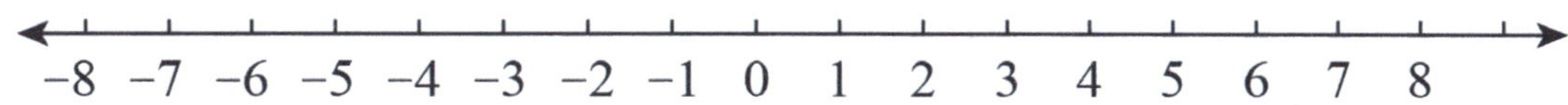

2 Solve the following absolute-value inequalities:

(1) $|2x - 3| > 5$

(2) $|\dfrac{2x + 1}{2}| \geq 5$

1 Solve the absolute-value inequality $|3x - 2| > 4$.

A. $x > 2$

B. $x < -\dfrac{2}{3}$

C. $x > 2$ or $x < -\dfrac{2}{3}$

D. $-\dfrac{2}{3} < x < 2$

2 Solve the absolute-value inequality $\left|\dfrac{3 - 2x}{3}\right| \geq 7$.

1 Solve the following absolute-value equations.
(1) $|4x + 8| = 12$.
$x = $ _______ or $x = $ _______ .

(2) $2\,|7 - 5x| - 13 = 3$.
$x = $ _______ or $x = $ _______ .

2 Solve the absolute-value equation: $\dfrac{3}{4}|x + 5| - 2 = 4$.
$x = $ _______ or $x = $ _______ .

3 The solution of $|2x - 4| > 7$ is _______ .

A. $x > \dfrac{11}{2}$

B. $x < -\dfrac{3}{2}$

C. $-\dfrac{3}{2} < x < \dfrac{11}{2}$

D. $x > \dfrac{11}{2}$ or $x < -\dfrac{3}{2}$

Summary:

Mind Map

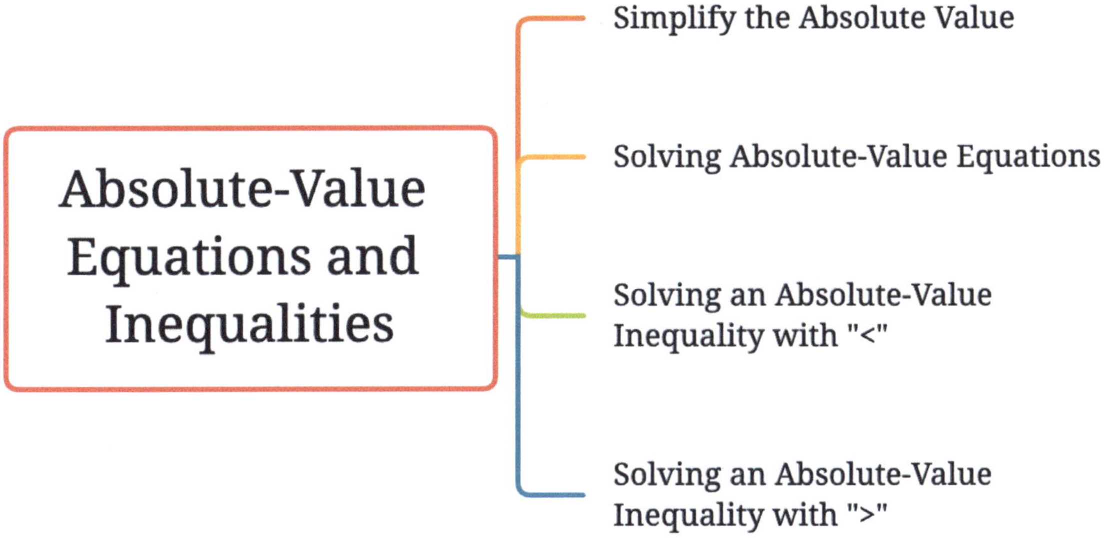
Absolute-Value Equations and Inequalities
Simplify the Absolute Value
Solving Absolute-Value Equations
Solving an Absolute-Value Inequality with "<"
Solving an Absolute-Value Inequality with ">"

1 Solve the following absolute-value equations.

$|7 - 5x| = 8$

$x =$ _______ or $x =$ _______ .

2 Solve the absolute-value equation: $4 - |3x - 4| = 2$.

$x =$ _______ or $x =$ _______ .

3 Solve the following absolute-value equation: $4 - 2\left|\dfrac{1}{2}x + 1\right| = 3$.

$x =$ ______ or $x =$ ______ .

4 The solution of $|x + 4| < 2$ is ______ .
A. $x > -2$ 　　　　　　　　　　　　　　　　B. $x < -6$
C. $x > -2$ or $x < -6$ 　　　　　　　　　　D. $-6 < x < -2$

5 Solve the following absolute-value inequalities.

(1) $|2x + 3| \leqslant 1$

A. $x \leqslant -2$

B. $x \geqslant -1$

C. $x \leqslant -2$ or $x \geqslant -1$

D. $-2 \leqslant x \leqslant -1$

(2) $|5 - 2x| \leqslant 2$

$$\underline{\hspace{2cm}} \leqslant x \leqslant \underline{\hspace{2cm}} \, .$$

6 Solve the following absolute-value inequalities.

(1) $|5 - x| \leqslant 2$

(2) $\left| \dfrac{x - 1}{2} \right| < 5$

7 The solution of $|x + 2| > 6$ is $x >$ _______ or $x <$ _______ .

8 Solve the following absolute-value inequalities.
(1) $|3x - 2| > 4$

    A. $x > 2$ B. $x < -\dfrac{2}{3}$

    C. $x > 2$ or $x < -\dfrac{2}{3}$ D. $-\dfrac{2}{3} < x < 2$

(2) $|5x + 2| \geqslant 3$
$x \geqslant$ _______ or $x \leqslant$ _______ .

9 Solve the following absolute-value inequalities.

(1) $\left|\dfrac{3x-4}{2}\right| > 4$

(2) $\left|\dfrac{2x-5}{3}\right| + 4 \geqslant 8$

10 Solve the following absolute-value inequalities.

(1) $\left|\dfrac{x-1}{3}\right| < 4$

A. $x < -11$

B. $x > 13$

C. $x < -11$ or $x > 13$

D. $-11 < x < 13$

(2) $\left|\dfrac{3x-4}{2}\right| + 4 \geqslant 9$

A. $x \leqslant -2$

B. $x \geqslant \dfrac{14}{3}$

C. $x \leqslant -2$ or $x \geqslant \dfrac{14}{3}$

D. $-2 \leqslant x \leqslant \dfrac{14}{3}$

2024 Fall Algebra 1B A Lesson 1-8

Lesson 2
Solving Systems of Linear Equations

1. Identify systems of linear equations in two variables.

2. Use substitution and elimination to solve systems of linear equations.

3. Solve word problems using systems of linear equations.

A system of linear equations in two variables is formed by two linear equations with the same two variables. Below is an example of a system of linear equations in two variables.

$$\begin{cases} x + y = 3 \\ x - 2y = 4 \end{cases}$$

Math Exploration 1

Which of the following systems is a linear system in two variables?

A. $\begin{cases} x - 3 = 0 \\ 3x - 2y = 7 \end{cases}$ 　　 B. $\begin{cases} 2x - y = 3 \\ 3xy = 8 \end{cases}$ 　　 C. $\begin{cases} x + y = 3 \\ x - z = 5 \end{cases}$ 　　 D. $\begin{cases} \dfrac{1}{2}x + \dfrac{3}{y} = 4 \\ \dfrac{1}{3}x + \dfrac{1}{2}y = 1 \end{cases}$

Practice 1

Which of the following systems is a linear system in two variables? _______ .

A. $\begin{cases} x - y = 1 \\ xy = 2 \end{cases}$ 　　 B. $\begin{cases} 4x - y = 1 \\ y = 2x + 3 \end{cases}$ 　　 C. $\begin{cases} x^2 - x - 2 = 0 \\ y = x + 1 \end{cases}$ 　　 D. $\begin{cases} \dfrac{1}{x} - 1 = y \\ 3x + y = 0 \end{cases}$

◎ ***Concept 2: Solving Systems of Linear Equations by Substitution***

A system of linear equations can be solved by the substitution method. The key points are expressing one of the variables with the other one, and then substituting the variable to solve the equation with only one variable.

Math Exploration 2

1 There is a system of linear equations $\begin{cases} x + 2y = 3 \ ① \\ 2x + 3y = 6 \ ② \end{cases}$. Which of the followings is a better way to express x in terms of y?

A. $x = \dfrac{6 - 3y}{2}$ B. $x = 6 - 3y$ C. $x = 3 - 2y$ D. $x = 2 - 3y$

Here are the steps for solving a linear system by this method.

1. Solve one of the equations for one of the variables.

2. Substitute the result from Step 1 in the other equation to find the value of the other variable.

3. Substitute the value from Step 2 in the revised equation from Step 1 to find the value of the remaining variable.

4. Write down the solution of the linear system.

Example:

Solve the linear system by substitution $\begin{cases} 2x - y = 1 \\ 3x + 2y = 12 \end{cases}$.

Step 1: Solve y with x.

From the first equation, we can have $y = $ _______ .

Step 2: Substitute y in the second equation to find the value of x.

$3x + 2$ _______ $= 12,$

$x = $ _______ .

Step 3: Substitute x in the equation from Step 1.

$y = $ _______ .

Step 4: Write down the solution of the linear system.

$\begin{cases} x = \\ y = \end{cases}$

2. Solve the linear system by substitution $\begin{cases} 3x - y = 5 \\ 5x + 2y = 23 \end{cases}$.

1 Given that $-3x + 4y = -1$, express y in terms of x. The answer is $y = $ ______ .

2 Solve the linear system by substitution: $\begin{cases} x - 2y = 3 \\ 3x + y = 2 \end{cases}$.

A linear system can also be solved by Elimination. We can eliminate one of the variables by adding or subtracting the two equations.

For example,

when we solve $\begin{cases} 2x + y = 4 \\ 2x - y = 2 \end{cases}$,

we can subtract the first equation by the second one, since we have the same $2x$ in the both two equations.

Therefore, after subtracting, we can eliminate x, then we have $2y = 2$,

thus, $y =$ _______ .

Then, we plug y into either equation to find x.

$x =$ _______ .

Sometimes, we do not have the same term in the two equations,

then we have to multiply the equations by some factors to get the same term,

then we can do the addition or subtraction.

Math Exploration 3

1 When solving the linear system $\begin{cases} 2x + 3y = 3 \\ 3x - 2y = 11 \end{cases}$ by elimination, we need to multiply and combine the equations to eliminate one variable. Which of the following conversions of the system is correct?

A. $\begin{cases} 4x + 6y = 3 \\ 9x - 6y = 11 \end{cases}$
B. $\begin{cases} 6x + 3y = 9 \\ 6x - 2y = 22 \end{cases}$
C. $\begin{cases} 4x + 6y = 6 \\ 9x - 6y = 33 \end{cases}$
D. $\begin{cases} 6x + 9y = 3 \\ 6x - 4y = 11 \end{cases}$

2 Solve the linear system by elimination: $\begin{cases} 7x + 4y = 2 \\ 3x - 6y = 24 \end{cases}$.

1. When solving the linear system $\begin{cases} 2x - 3y = 5 \text{①} \\ x = 3y + 7 \text{②} \end{cases}$ by linear combinations, we need to multiply and combine the equations to eliminate one variable. Which of the following is the correct sequence of procedures to eliminate one variable?

A. ①+② gives $3x = 12$

B. ①−② gives $x = -2$

C. ②×2−① gives $3y = 12$

D. ②−① gives $x = 2$

2. Solve the system of linear equations by elimination: $\begin{cases} 3x - 2y = -1 \\ 2x + 3y = 8 \end{cases}$.

Math Exploration 4

Formulae of going downstream and upstream:

The speed of the boat going downstream = The speed of the boat in still water + The speed of current

The speed of the boat going upstream = The speed of the boat in still water − The speed of current

A boat can travel 45 km downstream in three hours and travel 65 km upstream in five hours. Assume that the boat travels at a speed of x km/h in still water and the speed of the current is y km/h. What is the speed of the boat in still water and what is the speed of the current?

Practice 4

There are 3 more chickens than rabbits, and rabbits have 26 more legs than chickens have. How many chickens and rabbits are there? There are ______ chickens and ______ rabbits. (Hint: a chicken has two legs; a rabbit has four legs)

Exit Ticket

1 Solve the linear system by substitution $\begin{cases} 4x + 3y = 6 \\ 2x - y = 8 \end{cases}$.

$x =$ _______ , $y =$ _______ .

2 Solve the linear system by elimination: $\begin{cases} 2x - 5y = 17 \\ 3x + 2y = -3 \end{cases}$.

$x =$ _______ , $y =$ _______ .

Exit Ticket

3 The distance between City A and City B is 80 km. A boat can travel downstream from A to B in 4 hours and travel upstream from B to A in 5 hours, then the speed of the boat in still water is _______ km/h.

Mind Map

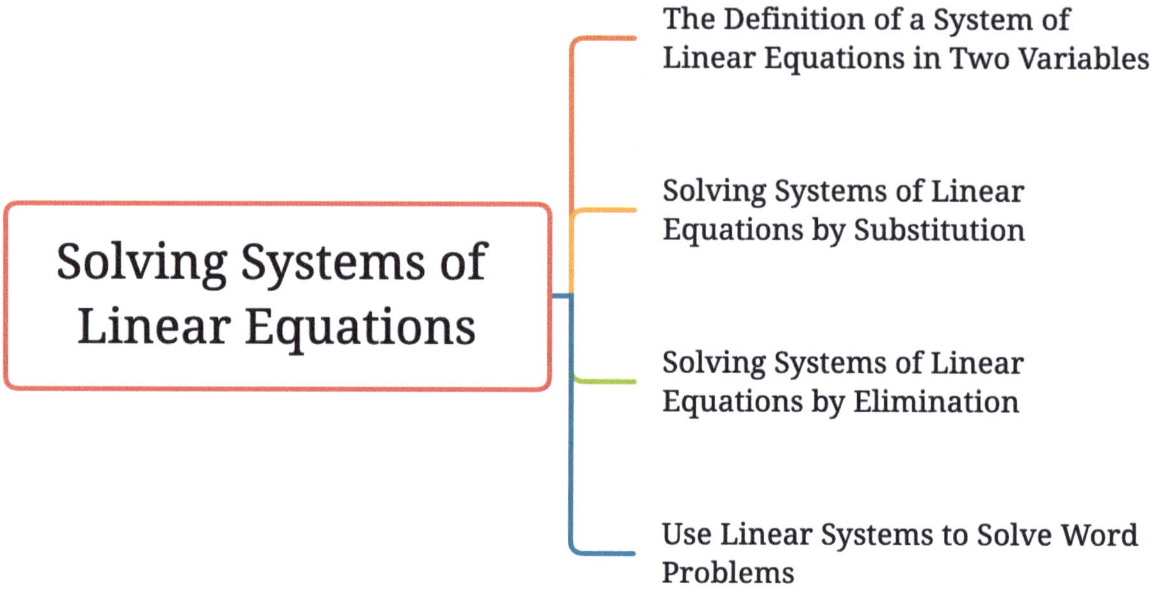
Solving Systems of Linear Equations
The Definition of a System of Linear Equations in Two Variables
Solving Systems of Linear Equations by Substitution
Solving Systems of Linear Equations by Elimination
Use Linear Systems to Solve Word Problems

1 Which of the following systems is a linear system in two variables?

A. $\begin{cases} x - y = 1 \\ x + z = 3 \end{cases}$

B. $\begin{cases} 2x + 3y = 5 \\ 5xy = 6 \end{cases}$

C. $\begin{cases} 3x - 1 = 0 \\ 5x + 3y - 5 = 0 \end{cases}$

D. $\begin{cases} \dfrac{3}{5}x + \dfrac{2}{y} = 6 \\ \dfrac{2}{3}x + 2y = 1 \end{cases}$

2 Which of the following systems is a linear system in two variables? _______ .

A. $\begin{cases} x - y = 4 \\ 3x - y^2 = 7 \end{cases}$

B. $\begin{cases} 3x + 5y = 10 \\ x^{-1} + y^{-1} = 2 \end{cases}$

C. $\begin{cases} x - 4y = 5 \\ 6x + y = 9 \end{cases}$

D. $\begin{cases} x - 1 = \dfrac{1}{y} \\ 3x + y = 7 \end{cases}$

3 There is a system of linear equations $\begin{cases} 3x + 4y = 4 \ \text{①} \\ 4x + 3y = 6 \ \text{②} \end{cases}$. Which of the followings is a better way to express x in terms of y?

A. $x = -2 - y$ B. $x = -2 + y$ C. $x = 2 + y$ D. $x = 2 - y$

4 There is a system of linear equations $\begin{cases} x + 2y = 3 \ \text{①} \\ 2x + 3y = 6 \ \text{②} \end{cases}$. Which of the followings is a better way to express x in terms of y?

A. $x = \dfrac{6 - 3y}{2}$ B. $x = 6 - 3y$ C. $x = 3 - 2y$ D. $x = 2 - 3y$

2024 Fall Algebra 1B A Lesson 1-8

5 Given that $2x - 6y = -1$, express x in terms of y. The answer is $x = $ _______ .

6 Solve the linear system by substitution $\begin{cases} x - 2y = -1 \\ 5x + 3y = 8 \end{cases}$, $x = $ _______ , $y = $ _______ .

7 When solving the linear system $\begin{cases} x - 4y - 1 = 0 \ \text{①} \\ 2x - 3y = 6 \ \text{②} \end{cases}$ by linear combinations, we need to multiply and combine the equations to eliminate one variable. Which of the following is the correct sequence of procedures to eliminate one variable?

A. ①+② gives $3x = 7$

B. ①−② gives $y = 7$

C. ②−① gives $x = 5$

D. ①×2−② gives $-5y = -4$

8 Solve the linear system by substitution: $\begin{cases} 4x + 3y = 2 \\ 3x - 2y = -7 \end{cases}$.

2024 Fall Algebra 1B A Lesson 1-8

 There are 5 more chickens than rabbits, and rabbits have 30 more legs than chickens have. How many chickens and rabbits are there? There are ________ chickens and _______ rabbits. (Hint: a chicken has two legs; a rabbit has four legs)

10 The distance between City A and City B is 60 km. A boat can travel downstream from A to B in 3 hours and travel upstream from B to A in 5 hours, then the speed of current is ______ km/h.

Lesson 3
Inequalities and Systems of Linear Inequalities

1. Solve linear inequalities involving denominators and parentheses.

2. Solve compound inequalities involving "or", "and".

3. Convert the compound inequalities to the systems of linear inequalities.

1. Addition Property of Inequality

If $a > b$, then $a + c > b + c$ (a, b and c are all real numbers).

If $a < b$, then $a + c < b + c$ (a, b and c are all real numbers).

2. Subtraction Property of Inequality

If $a > b$, then $a - c > b - c$ (a, b and c are all real numbers).

If $a < b$, then $a - c < b - c$ (a, b and c are all real numbers).

3. Multiplication Property of Inequality

When $c > 0$,

if $a > b$, then $ac > bc$ (a, b and c are all real numbers);

if $a < b$, then $ac < bc$ (a, b and c are all real numbers).

When $c < 0$,

if $a > b$, then $ac < bc$ (a, b and c are all real numbers);

if $a < b$, then $ac > bc$ (a, b and c are all real numbers).

4. Division Property of Inequality

When $c > 0$,

if $a > b$, then $\dfrac{a}{c} > \dfrac{b}{c}$ (a, b and c are all real numbers);

if $a < b$, then $\dfrac{a}{c} < \dfrac{b}{c}$ (a, b and c are all real numbers).

When $c < 0$,

if $a > b$, then $\dfrac{a}{c} < \dfrac{b}{c}$ (a, b and c are all real numbers);

if $a < b$, then $\dfrac{a}{c} > \dfrac{b}{c}$ (a, b and c are all real numbers).

In conclusion,

when we add or subtract the same number on both sides, the inequality still holds;

when we multiply or divide the same positive number on both sides, the inequality still holds;

when we multiply or divide the same negative number on both sides, the inequality sign should be flipped.

1. If $a > b$, which of the following inequalities is not always correct?

    A. $a + c^2 > b + c^2$ B. $a - c^2 > b - c^2$ C. $ac^2 > bc^2$ D. $\dfrac{a}{c^2 + 1} > \dfrac{b}{c^2 + 1}$

2. Find the solution of the following inequalities.

    (1) $x + 2 > 3$ (2) $-3x < 9$

    (3) $-\dfrac{x}{4} \geqslant 1$

1 Which of the following statements is not correct?

A. If $a > b$, then $a + c > b + c$

B. If $a + c > b + c$, then $a > b$

C. If $a > b$, then $ac^2 > bc^2$

D. If $ac^2 > bc^2$, then $a > b$

2 Find the solution of the following inequalities.

(1) $x - 2 \leqslant -4$

A. $x \geqslant 2$

B. $x \geqslant -2$

C. $x \leqslant 2$

D. $x \leqslant -2$

(2) $-4x > -20$

A. $x < -5$

B. $x > -5$

C. $x < 5$

D. $x > 5$

(3) $-\dfrac{x}{3} \leqslant 2$

A. $x \geqslant 6$

B. $x \geqslant -6$

C. $x \leqslant 6$

D. $x \leqslant -6$

Linear inequality can be expressed as $kx + b > m$ ($k \neq 0$ and b, m are all real numbers).

1. Solving a linear inequality:

a. Eliminate the denominators

b. Remove the parentheses

c. Move the terms

d. Combine like terms

e. Change the coefficient of the variable to 1

Example:

Solve the inequality $\dfrac{2x + 4}{4} - x > 2.$

Step 1: Eliminate the denominators.

$2x + 4 - 4x > 8$

Step 2: Remove the parentheses.

There is no parentheses.

Step 3: Move the terms.

$2x - 4x > 8 - 4.$

Step 4: Combine like terms.

$-2x > 4.$

Step 5: Change the coefficient of the variable to 1.

$x < -2.$

2. Graphing an inequality:

We can use a number line to show an inequality.

For example,

the graph of $x > 1$:

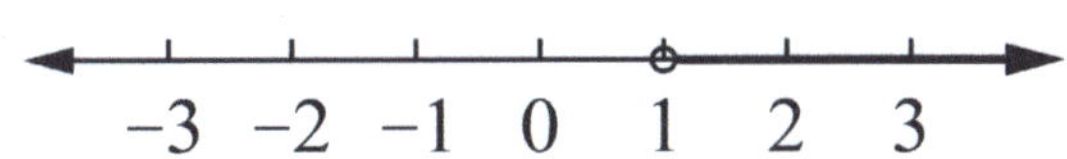

the graph of $x < 1$:

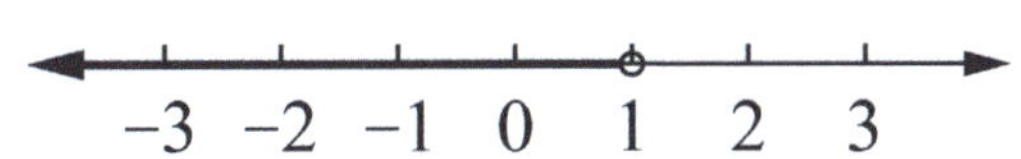

the graph of $x \geqslant 1$:

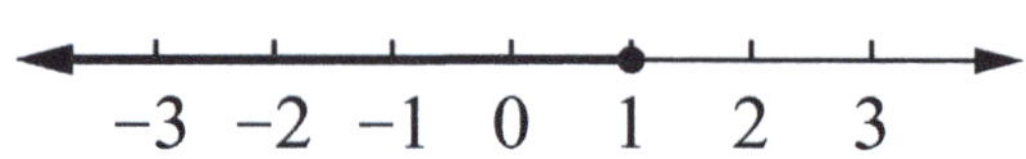

the graph of $x \leqslant 1$:

Math Exploration 2

Solve the following inequalities and graph each solution.

(1) $-3x \geqslant 2x - 15$

(2) $-2(3x + 2) \leqslant 4x - 10$

Practice 2

1 Solve the inequality $2 - 3x \geqslant 2x - 8$.

   A. $x \geqslant 2$ B. $x \leqslant 2$ C. $x \geqslant -2$ D. $x \leqslant -2$

2 Solve the inequality, and graph the solution.
$4(x - 1) + 3 \geqslant 3x$

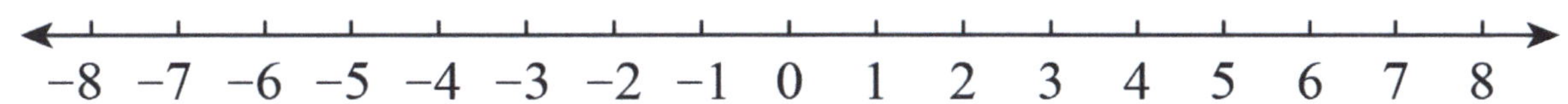

3 Solve the inequality $\dfrac{x - 3}{2} - 1 > x - 5$, and graph the solution.

For example, solving the compound inequalities $x - 1 < 0$ or $x + 1 > 4$.

The solution should be $x < 1$ or $x > 3$.

The graph of the compound inequalities is shown below. There are two parts in the graph, and the solution should involve both two parts.

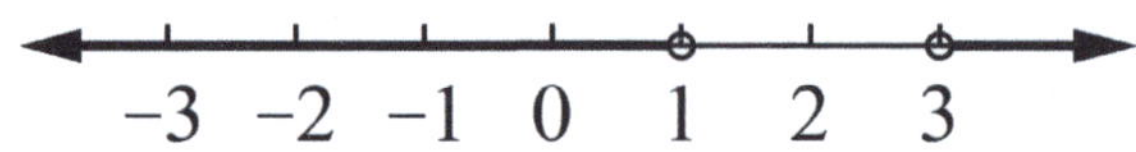

Math Exploration 3

Solve the inequalities and graph the solution.
$7x < -42$ or $x + 5 \geqslant 3$

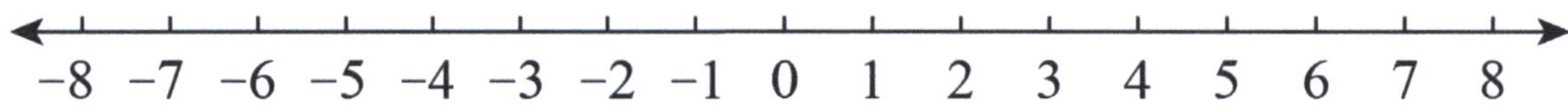

Practice 3

Solve the inequalities, and graph the solution.
$2x + 7 < -11$ or $-3x - 2 < 13$

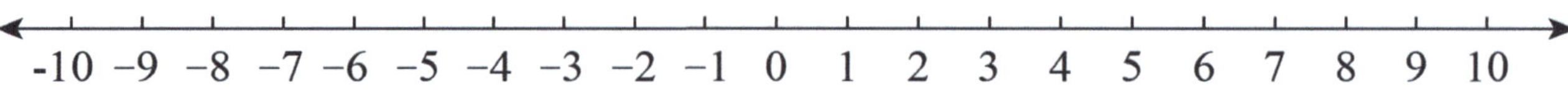

For example, solving the compound inequalities $2 < 2x + 4 < 6$.

This inequality can be divided into two parts which are $2 < 2x + 4$ and $2x + 4 < 6$, and the solution should satisfy both two inequalities.

Therefore, the solution is $-1 < x < 1$ and the graph of the compound inequalities is shown below.

$$\xleftarrow{\quad\quad} \underset{-3 \quad -2 \quad -1 \quad 0 \quad 1 \quad 2 \quad 3}{\rule{0pt}{0pt}} \xrightarrow{\quad\quad}$$

Moreover, we can also use the system of linear inequalities $\begin{cases} 2x + 4 > 2 \\ 2x + 4 < 6 \end{cases}$ to express the compound inequalities $2 < 2x + 4 < 6$.

Math Exploration 4

1 Solve the compound inequalities $-1 < \dfrac{1 - 2x}{4} \leqslant 2$ and graph the solution on the number line.

$$\xleftarrow{\quad\quad} \underset{-8 \enspace -7 \enspace -6 \enspace -5 \enspace -4 \enspace -3 \enspace -2 \enspace -1 \enspace 0 \enspace 1 \enspace 2 \enspace 3 \enspace 4 \enspace 5 \enspace 6 \enspace 7 \enspace 8}{\rule{0pt}{0pt}} \xrightarrow{\quad\quad}$$

2 Solve the system of linear inequalities: $\begin{cases} 3x - 1 > x + 1 \\ \frac{4x-5}{3} \leqslant x \end{cases}$.

1 Solve the inequalities: $x - 1 < 2 - 2x \leqslant x + 1$.

2 The solution of $\begin{cases} 2 - x < 1 \\ 5x < 3(x-2) \end{cases}$ is ______ .

A. $x > 1$ B. $-3 < x < 1$ C. $x > -3$ D. No solution

3 Solve the system of linear inequalities: $\begin{cases} x + 5 < 0 \\ \frac{3x-1}{2} > 2x + 1 \end{cases}$.

Exit Ticket

1 Solve the inequalities $-1 < \dfrac{2 - 2x}{2} < 2$.

________ $< x <$ ________ .

2 The solution of the system of linear inequalities $\begin{cases} 3x + 1 \geqslant 7 \\ 4x - 3 < 9 \end{cases}$ is ________ .

A. $x \geqslant 2$ B. $x < 3$ C. $x < \dfrac{3}{2}$ D. $2 \leqslant x < 3$

Summary:

Mind Map

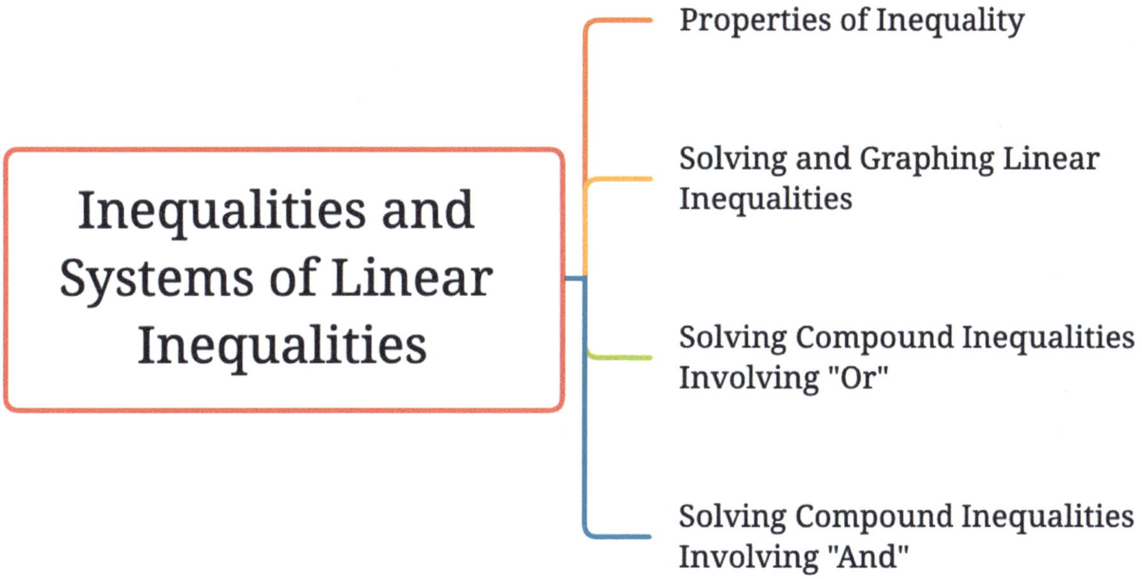
Inequalities and Systems of Linear Inequalities
Properties of Inequality
Solving and Graphing Linear Inequalities
Solving Compound Inequalities Involving "Or"
Solving Compound Inequalities Involving "And"

1 If $a > b$, $c < 0$, which of the following inequalities is not correct?

   A. $a + c > b + c$ B. $a - c^2 > b - c^2$ C. $ac < bc$ D. $\dfrac{a}{c} > \dfrac{b}{c}$

2 Find the solution of the following inequalities.

(1) $x - 5 > 2$.

   A. $x > 7$ B. $x > -3$ C. $x < 7$ D. $x < -3$

(2) $2x < -3$.

   A. $x > -\dfrac{3}{2}$ B. $x < -\dfrac{3}{2}$ C. $x > -\dfrac{2}{3}$ D. $x < -\dfrac{2}{3}$

(3) $-\dfrac{2x}{3} \leqslant 5$.

   A. $x \geqslant \dfrac{15}{2}$ B. $x \leqslant \dfrac{15}{2}$ C. $x \geqslant -\dfrac{15}{2}$ D. $x \leqslant -\dfrac{15}{2}$

3 Find the solution of the following inequalities.

(1) $-x + 3 \geqslant 2$

    A. $x \geqslant -1$ B. $x \geqslant 1$ C. $x \leqslant 1$ D. $x \leqslant -1$

(2) $-5x > -10$

    A. $x < -2$ B. $x > -2$ C. $x < 2$ D. $x > 2$

4 Solve the following inequalities and graph each solution.

(1) $-x - 2 \leqslant -3x + 4$

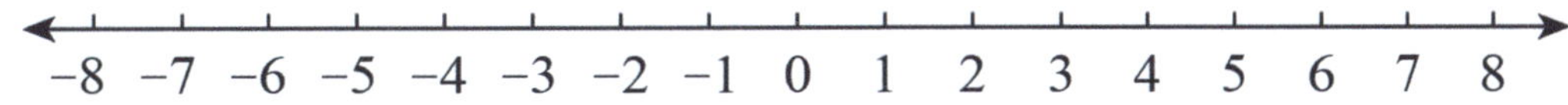

(2) $-2(3x + 1) > -5x$

5 Solve the following inequalities.

(1) $2x \geqslant -3x - 5$.

    A. $x \geqslant 5$ B. $x \leqslant 5$ C. $x \geqslant -1$ D. $x \leqslant -1$

(2) $-4(x - 1) \leqslant 2x - 1$.

    A. $x \leqslant \dfrac{5}{6}$ B. $x \geqslant \dfrac{5}{6}$ C. $x \geqslant -\dfrac{5}{6}$ D. $x \geqslant -\dfrac{5}{6}$

6 Solve the inequality $\dfrac{x + 2}{3} + 1 > -x + 2$, and graph the solution.

2024 Fall Algebra 1B A Lesson 1-8

7 Solve the inequalities and graph the solution.
$6x < -24$ or $x - 2 \geqslant 3$

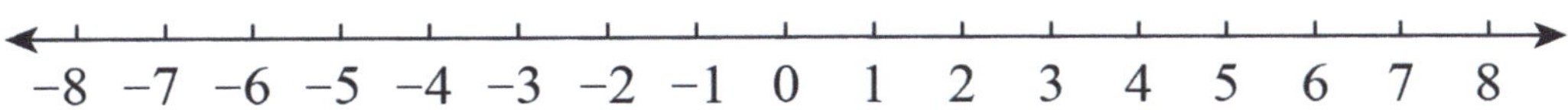

8 The solution of $\begin{cases} 3 - x < 1 \\ 4x < 3(x+1) \end{cases}$ is ______ .

A. $x > 2$ B. $2 < x < 3$ C. $x < 3$ D. No solution

9 Solve the inequalities $-2 < \dfrac{3 - 2x}{3} < 2$.

______ $< x <$ ______ .

10 The solution of the system of linear inequalities $\begin{cases} 2x + 3 \geqslant -3 \\ 5x - 3 < 7 \end{cases}$ is ______ .

    A. $x \geqslant -3$ B. $x < 2$ C. $x < -3$ D. $-3 \leqslant x < 2$

Lesson 4
Basics of Linear Equations

Learning Objectives

1. Understand the definition of direct variation and know how to write direct variation and how to read the graph of direct variation.

2. Understand the definition and graph of linear equations.

3. Use the equation to find the intersections.

3. Write the linear equations in point-slope form, slope-intercept form and standard form.

A linear equation is an algebraic equation where the highest degree of the variable in the given equation is 1. Linear equations can be written in the form of $y = mx + b$ ($m \neq 0$, m, b are constant values). We can use graphs to represent the linear equations. The graph of a linear equation is a line. m is the slope of the line and b is the y-intercept (the y-coordinate where the line intersects the y-axis) of the line.

The equation of direct variation is a special form of linear equation.

Math Exploration 1

Which of the followings is a linear equation?

A. $y = \dfrac{5}{x}$ B. $y = -5x + 3$ C. $y = x^2 + 3x - 5$ D. $y = \sqrt{2x + 4}$

Practice 1

Among the following equations: (1) $y = 3x$; (2) $y = 2x - 1$; (3) $y = \dfrac{1}{x}$; (4) $y = x^2 - 1$; (5) $y = \dfrac{x}{8}$, how many of them is(are) linear equation(s)?

A. 4 B. 3 C. 2 D. 1

◎ *Concept 2: Properties of Linear Equations*

Linear Equation	$y = mx + b$ $(m \neq 0)$					
Feature of Graph	Line passes through point $(0, b)$					
m, b	$m > 0$			$m < 0$		
	$b > 0$	$b < 0$	$b = 0$	$b > 0$	$b < 0$	$b = 0$
Graph						
Quadrant	Passes through Quadrants I, III for certain			Passes through Quadrants II, IV for certain		
Feature of Graph	The graph goes up from left to right			The graph goes down from left to right		
Property	y increases as x increases			y decreases as x increases		

Math Exploration 2

1 Given that both points $(-1, y_1)$ and $(2, y_2)$ are on the linear equation $y = -3x + b$, what is the relationship between y_1 and y_2?

  A. $y_1 > y_2$ B. $y_1 = y_2$ C. $y_1 < y_2$

2 Given that the linear equation is $y = -3x + 1$, which of the following statements is true?

A. Its graph passes through point $(-1, 3)$.

B. y increases as x increases.

C. When $x > 0$, $y < 1$.

D. Its graph passes through Quadrants I, II and III.

Practice 2

1 Given that both points $P_1(x_1, y_1)$, $P_2(x_2, y_2)$ are on line $y = -4x + 3$, and $x_1 < x_2$, what is the relationship between y_1 and y_2?

A. $y_1 > y_2$ B. $y_1 > y_2 > 0$ C. $y_1 < y_2$ D. $y_1 = y_2$

 Which of the following statements is correct about line $y = -3x + 1$?

A. The point $(-1, 3)$ lies on the line.

B. y increases as x increases.

C. The line passes through Quadrants I, II and III.

D. The line meets the x-axis and the y-axis at $\left(\frac{1}{3}, 0\right)$, $(0, 1)$, respectively.

1. Intersections with Two Axes:

To find the coordinates of a point given a linear equation and one axis (either the x-axis or the y-axis), substitute the x or y value into 0 and use the equation to find the other coordinate.

2. Intersections of Two Linear Equations:

To find the intersection of two linear equations, set the two equations equal to each other and solve for the variables. The solution will give you the coordinates of the point where the two lines intersect.

Math Exploration 3

(1) Line $y = -2x - \dfrac{1}{2}$ meets the x-axis at ______ , and the y-axis at ______ .

(2) Lines $y = x + 1$ and $y = -2x + 2$ meet at point ______ .

1 Line $y = 2x + 10$ intersects the x-axis at _____ .
   A. $(5, 0)$ B. $(0, 5)$ C. $(-5, 0)$ D. $(0, -5)$

2 (1) Lines $y = -2x + 3$ and $y = x - 1$ meet at point (_____ , _____).

(2) Lines $y = -x$ and $y = 5x - 6$ meet at point (_____ , _____).

We can write the equation of a line in different forms.

(1) If we know the slope of the line m and its y-intercept b, we can write the equation in the slope-intercept form $y = mx + b$.

If the line passes through points (x_1, y_1) and (x_2, y_2), we can get the slope: $m = \dfrac{y_2 - y_1}{x_2 - x_1}$.

(2) If we know the line passes through point (x_1, y_1) and its slope m, we can write the equation in the point-slope form $y - y_1 = m(x - x_1)$.

(3) The standard form of a linear equation is $Ax + By = C$, where A, B, and C are integers, $A > 0$, and both A and B are not zero.

Math Exploration 4

1 Given that the slope of the line is -2 and it passes through point $(1, 1)$, write the equation of it in point-slope form and convert it to the slope-intercept form.

2 The graph of the line is shown below, write the equation of it in slope-intercept form.

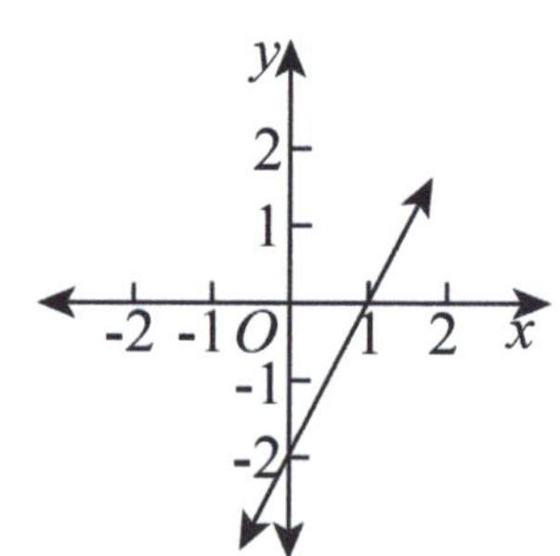

1. Given that the slope of the line is 2 and it passes through points $(-4, -9)$, write down the equation of line l_1 in point-intercept form and convert it to the slope-intercept form.

2. If the line passes through points $(1, 2)$ and $(-2, 3)$, find the equation of this line.

Math Exploration 5

Write the linear equations $y = \dfrac{1}{2}x + 2$ and $y = -\dfrac{4}{3}x - 1$ in the standard form with integer coefficients.

(1) Write the standard form of the line with integer coefficients passing through $(1, 2)$ and $(-3, 3)$.

(2) Write the standard form of the line with integer coefficients passing through $(1, 2)$ with a slope of 3.

1 Given that both points $A(x_1, y_1)$, $B(x_2, y_2)$ are on the graph of the direct variation $y = -x$ and $x_1 < x_2$, then ______ .

A. $y_1 > y_2$ B. $y_1 < y_2$ C. $y_1 = y_2$

2 Line l passes through $(2, 4)$, and its y-intercept is 8. The equation of line l is $y =$ ______ .

Summary:

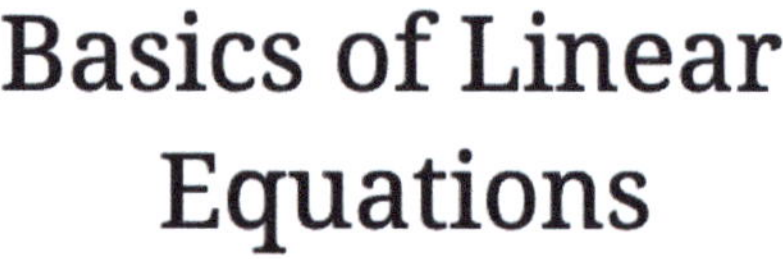

Basics of Linear Equations

- Concept of Linear Equations
- Properties of Linear Equations
- Intersections of Linear Equations
- Different Forms of Linear Equations

1 Given that a linear equation is $y = mx + (1 - m)$, if $0 < m < 1$, it will not pass through Quadrant _______ .

   A. *I*   B. *II*   C. *III*   D. *IV*

2 Given that both points $(-1, y_1)$ and $(2, y_2)$ are on the linear equation $y = 2x + b$, what is the relationship between y_1 and y_2?

   A. $y_1 > y_2$ B. $y_1 = y_2$ C. $y_1 < y_2$

3 The graph of the linear equation $y = mx + b$ is shown in the figure below. Which of the following statements is true?

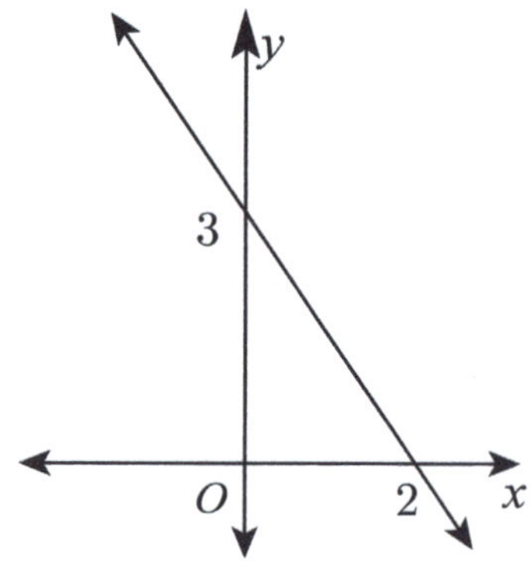

A. The line intersects the y-axis at $(3, 0)$.
B. y increases as x increases.
C. When $0 \leqslant x < 2$, $0 < y \leqslant 3$.

4 Which of the following statements is incorrect about line $y = mx + m \ (m \neq 0)$?
A. Point $(0, m)$ is on the line.
B. The line passes through a fixed point $(-1, 0)$.
C. When $m > 0$, y increases as x increases.
D. The line will pass through Quadrants I, II, III.

5 Line $y = x + \dfrac{3}{2}$ meets the x-axis at (_____ , _____) and the y-axis at (_____ , _____).

6 (1) Lines $y = -x + 2$ and $y = x + 4$ meet at point _____ .

(2) Lines $y = -2x + 1$ and $y = -x - 1$ meet at point _____ .

7 (1) Given that the slope of the line is -5 and its y-intercept is 1, write the equation of it in slope-intercept form.

(2) Given that the slope of the line is -3 and it passes through point $(1, 2)$, write the equation of it in point-slope form and convert it to the slope-intercept form.

 The graphs of two lines are shown below, write the equations of them in slope-intercept form.

(1)

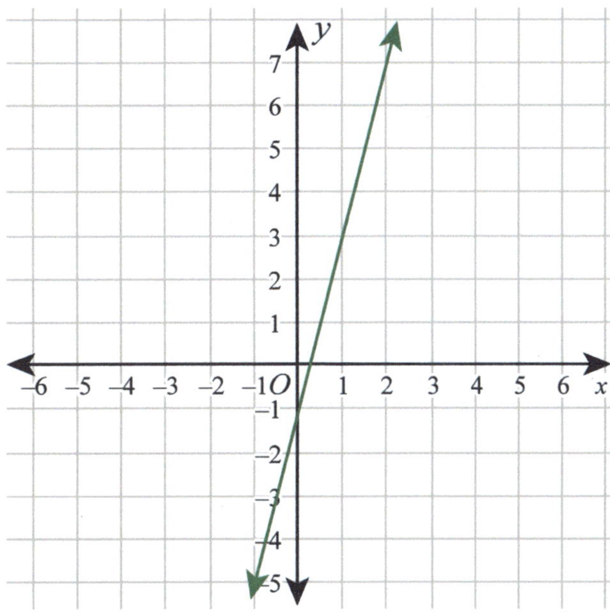

(2)

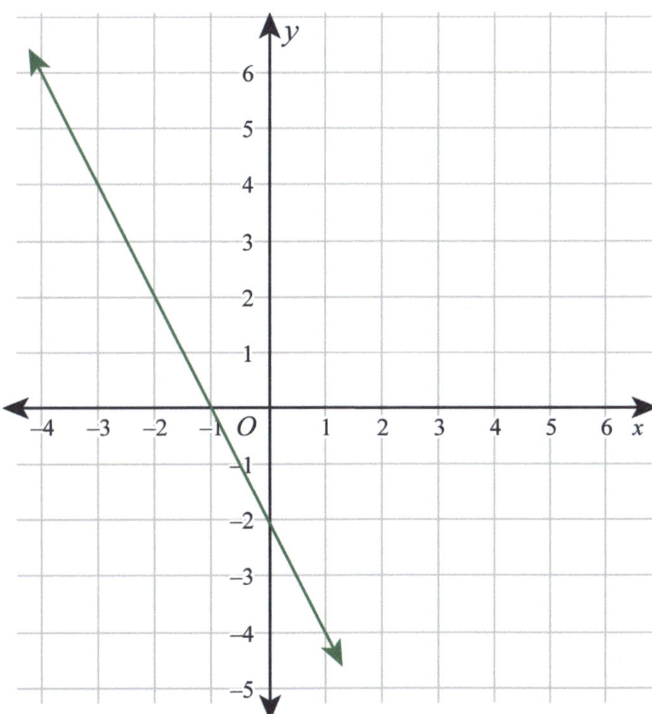

 Find the equations of the following lines.
(1) Line l_1 passes through $(1, -2)$, and its y-intercept is 2. Find the equation of line l_1.

(2) Line l_2 passes through points $(-1, 1)$ and $(3, 6)$. Find the equation of line l_2.

10 (1) Write a standard form of the line with integer coefficients passing through $(-2, 1)$ and $(2, 3)$.

(2) Write a standard form of the line with integer coefficients passing through $(3, 1)$ with a slope of -2.

2024 Fall Algebra 1B A Lesson 1-8

Lesson 5
Transformations of Linear Equations

1. Use slopes to determine the relationship between two lines.

2. Understand how the equation changes when the graph is translated and reflected.

In geometry, the relationship between two lines can be categorized as parallel or perpendicular.

1. **Parallel lines** are two or more lines that never intersect. They are always equidistant and maintain the same direction. In other words, they have the same slope but different y-intercepts. Symbolically, if we have two lines with equations $y = m_1 x + b_1$ and $y = m_2 x + b_2$, then the lines are parallel if and only if $m_1 = m_2$.

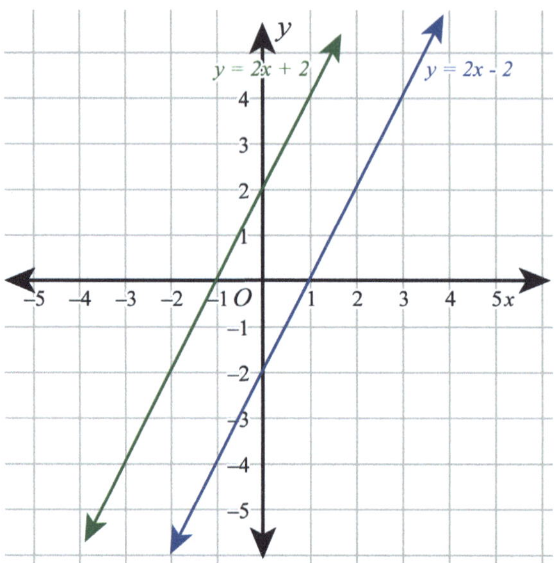

2. **Perpendicular lines** are two that intersect at a 90-degree angle. The slopes of perpendicular non-vertical lines are negative reciprocals of each other. If we have two lines with slopes m_1 and m_2, then they are perpendicular if and only if $m_1 \times m_2 = -1$.

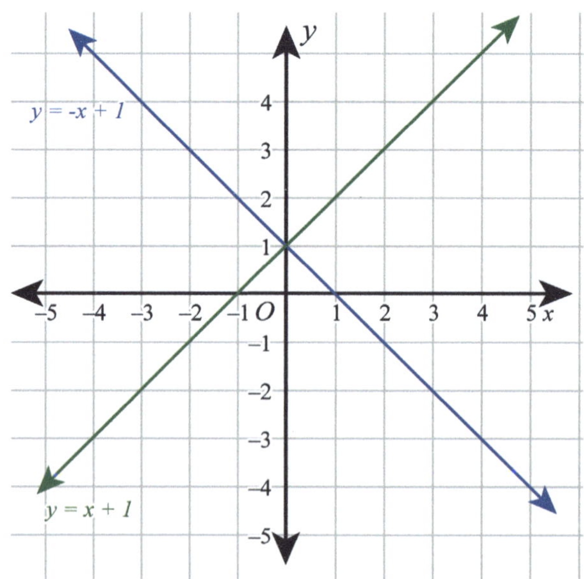

Answer the following questions.

(1) A line l that passes through point $(0, 0)$ is parallel to line $y = 3x + 4$. What is the equation of line l?

(2) A line l that passes through point $(0, 0)$ is perpendicular to line $y = 3x + 4$. What is the equation of line l?

Answer the following questions.

(1) A line l that passes through point $(2, 3)$ is parallel to line $y = 2x + 2$. What is the equation of line l?

(2) A line l that passes through point $(2, 3)$ is perpendicular to line $y = -\dfrac{1}{2}x + 4$. What is the equation of line l?

◎ *Concept 2: Translation*

If a linear equation $y = mx + b$ moves up g unit(s), the equation will be $y = mx + b + g$.

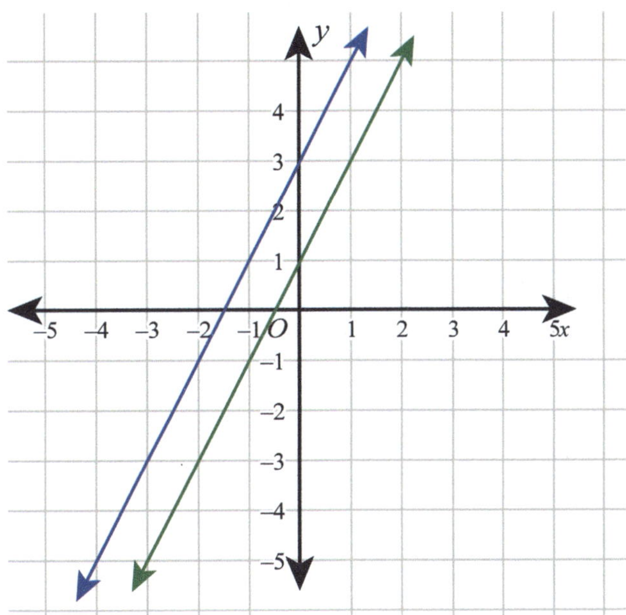

If a linear equation $y = mx + b$ moves down h unit(s), the equation will be $y = mx + b - h$.

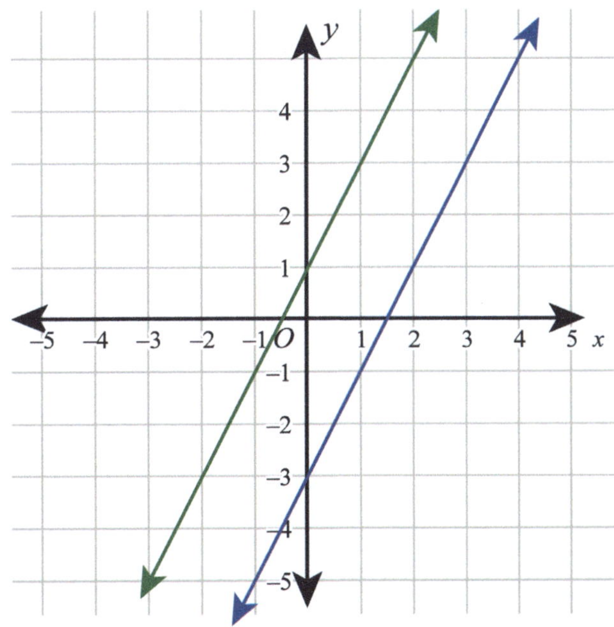

If a linear equation $y = mx + b$ moves p unit(s) to the left, the equation will be $y = m(x + p) + b$.

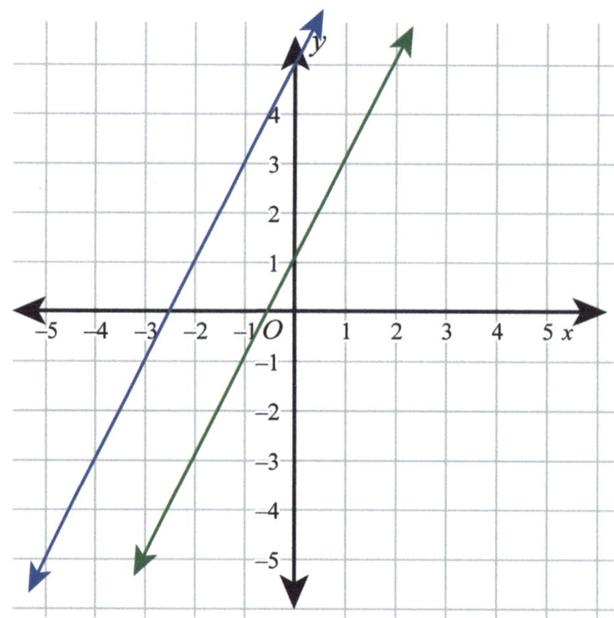

If a linear equation $y = mx + b$ moves q unit(s) to the right, the equation will be $y = m(x - q) + b$.

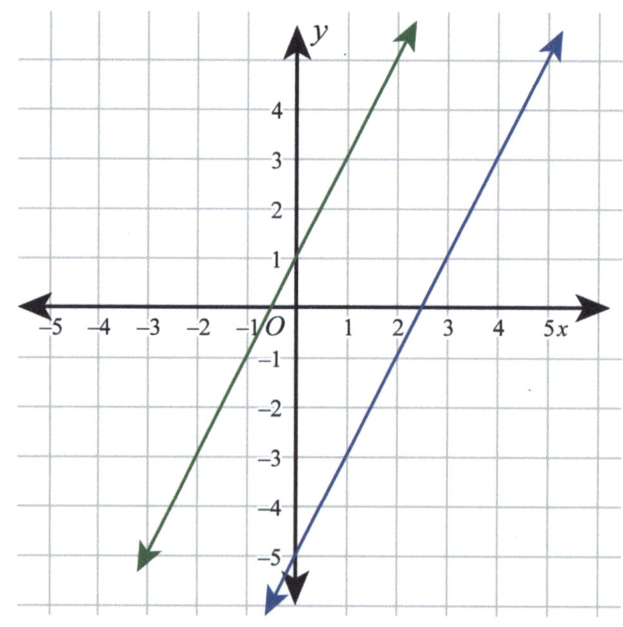

1 Answer the following questions.

(1) If the graph of the linear equation $y = \frac{1}{2}x - 1$ moves down 2 units, the equation will be _______ .

(2) If the graph of the linear equation $y = \frac{1}{2}x - 1$ moves up 3 units, the equation will be _______ .

(3) If the graph of the linear equation $y = \frac{1}{2}x - 1$ moves 4 units to the right, the equation will be _______ .

(4) If the graph of the linear equation $y = \frac{1}{2}x - 1$ moves 6 units to the left, the equation will be _______ .

2 Line l_1 first moves 3 units to the right and then moves down 2 units, and it becomes line l_2 whose equation is $y = 2x - 1$. The equation of line l_1 is _______ .
A. $y = 2x + 7$ B. $y = -2x + 7$ C. $y = 2x - 7$ D. $y = -2x - 7$

1. If the graph of the linear equation $y = -2x - 2$ first moves 1 units to the right and then moves up 3 units, the equation will be _______ .
   A. $y = 2x - 1$ B. $y = -2x + 1$ C. $y = -2x$ D. $y = -2x + 3$

2. If the graph of the linear equation $y = 2x + 3$ first moves 3 units to the left and then moves down 2 units, the equation will be _______ .
   A. $y = -2x + 7$ B. $y = 2x + 7$ C. $y = -2x - 7$ D. $y = 2x - 7$

If line l and line $y = mx + b$ are symmetric about the origin, the equation of line l is $y = mx - b$.

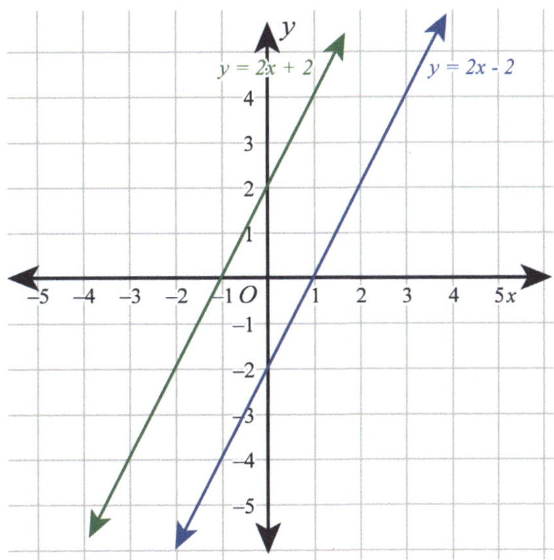

Math Exploration 3

1. Line $y = mx + b$ and line $y = -3x + 7$ are symmetric about the origin, so $m =$ _______ , $b =$ ______ .

2. In a coordinate plane, if line l and line $y = -\dfrac{1}{2}x + 2$ are symmetric about the origin, and line l passes through point $(2, m)$, then the value of m is ______ .
   A. -1 B. -2 C. -3 D. -4

1 Line l and line $y = 2x + 1$ are symmetric about the origin. The equation of line l is ______ .

A. $y = 2x + 1$ B. $y = 2x - 1$ C. $y = -2x + 1$ D. $y = -2x - 1$

2 In a coordinate plane, line l and line $y = 2x - 3$ are symmetric about the origin, and line l passes through point $(5, m)$, so the value of m is ______ .

1. Symmetry about the x-axis:

The equation of the line symmetry of the linear equation $y = mx + b$ across the x-axis is $y = -mx - b$.

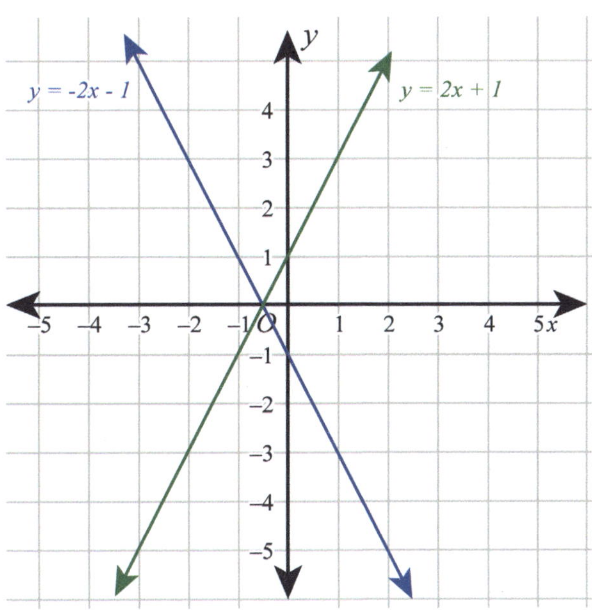

2. Symmetry about the y-axis:

The equation of the line symmetry of the linear equation $y = mx + b$ across the y-axis is $y = -mx + b$.

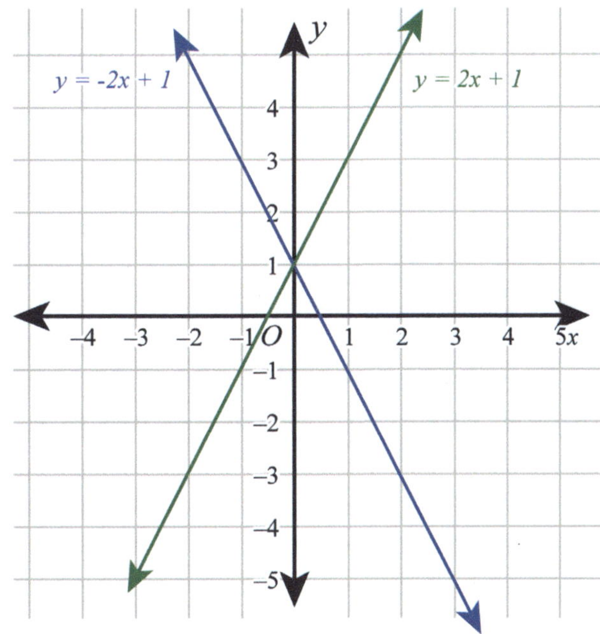

(1) If line l_1 and line $y = 2x + 3$ are symmetric about the y-axis, the equation of line l_1 is ______ .

(2) If line l_2 and the line $y = -3x + 5$ are symmetric about x-axis, the equation of line l_2 is ______ .

Answer the following questions.
(1) If line l_1 and line $y = -x + 2$ are symmetric about the y-axis, the equation of line l_1 is ______ .

(2) If line l_2 and line $y = 2x - 3$ are symmetric about the y-axis, and line l_3 and line l_2 are symmetric about the x-axis, the equation of line l_3 is ______ .

1 Line l is perpendicular to line $y = x + 1$ and passes through $(2, 3)$, then the equation of line l is ______ .

2 Move down line $y = 2x$ by three units. Now the line will pass point $(2, a)$, then $a =$ ______ .

A. 2   B. 1   C. 3   D. 0.5

3 If line l and line $y = -2x - 1$ are symmetric about the origin, then the equation of line l is ______ .

A. $y = -2x + 1$ B. $y = -2x - 1$ C. $y = 2x - 1$ D. $y = 2x + 1$

Summary:

2024 Fall Algebra 1B A Lesson 1-8

Mind Map

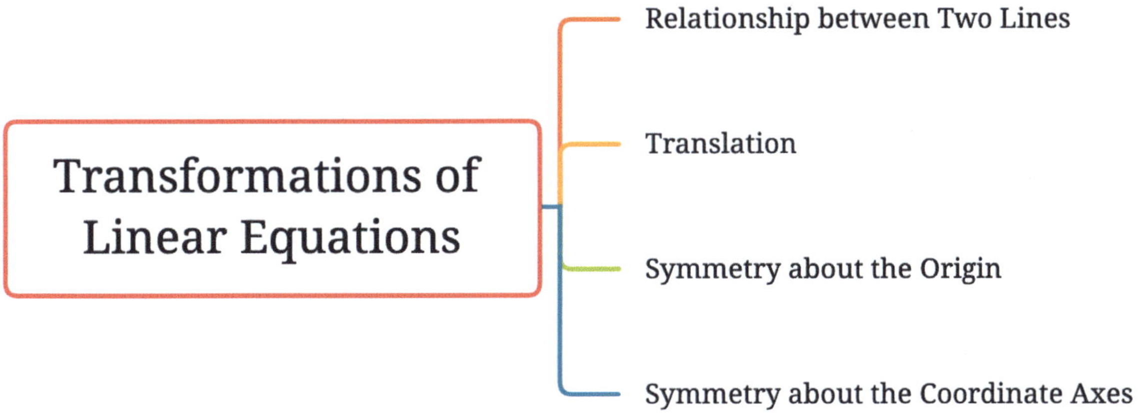
Transformations of Linear Equations
Relationship between Two Lines
Translation
Symmetry about the Origin
Symmetry about the Coordinate Axes

1 **(1)** If line l_1: $y = mx + b$ passes through point $(4, 7)$ and is perpendicular to line $y = 2x + 3$, what is the equation of line l_1?

(2) If line l_2: $y = mx + b$ passes through point $(-3, 4)$ and is perpendicular to the line $y = -3x - 1$, what is the equation of line l_2?

2 **(1)** If line l_1: $y = mx + b$ passes through point $(2, 3)$ and is perpendicular to line $y = \dfrac{1}{2}x - 5$, what is the equation of line l_1?

(2) If line l_2: $y = mx + b$ passes through point $(-3, -1)$ and is perpendicular to line $y = -2x + 4$, what is the equation of line l_2?

2024 Fall Algebra 1B A Lesson 1-8

3 A line passing through the points $A(-2, m)$ and $B(m, 4)$ is perpendicular to the line $2x + y - 1 = 0$, then $m =$ _______ .

A. 0   B. −8   C. 2   D. 10

4 Answer the following questions.

(1) If the graph of the linear equation $y = 3x - 2$ moves down 4 units, the equation will be _______ .

(2) If the graph of the linear equation $y = 3x - 2$ moves up 5 units, the equation will be _______ .

(3) If the graph of the linear equation $y = 3x - 2$ moves 3 units to the right, the equation will be _______ .

(4) If the graph of the linear equation $y = 3x - 2$ moves 2 units to the left, the equation will be _______ .

5 If the graph of the linear equation $y = -x - 2$ first moves 2 units to the right and then moves up 4 units, the equation will be ______ .
A. $y = x + 4$ 　　　B. $y = -x - 4$ 　　　C. $y = -x + 4$ 　　　D. $y = x - 4$

6 Line l_1 and line $y = -2x + 5$ are symmetric about the origin, so the equation of line l_1 is ______ .

7 In a coordinate plane, line l and line $y = 4x - 1$ are symmetric about the origin, and line l passes through point $(2, m)$, so the value of m is ______ .

2024 Fall Algebra 1B A Lesson 1-8

8 (1) If line l_1 and line $y = -5x + 7$ are symmetric about the y-axis, the equation of line l_1 is _______ .

(2) If line l_2 and line $y = 4x - 3$ are symmetric about the x-axis, the equation of line l_2 is _______ .

9 Line $y = mx + n$ and line $y = -4x - 3$ are symmetric about the origin, so $m =$ _______ , $n =$ _______ .

10 In a coordinate plane, line l and line $y = \dfrac{1}{3}x - 4$ are symmetric about the origin, and line l passes through point $(6, m)$, then the value of m is ______ .

Lesson 6
Applications of Linear Equations

Learning Objectives

1. Understand the meaning of the intersection of the graphs of linear equations.

2. Use the graphs of linear equations to solve the linear system.

3. Use the graph to determine the number of solutions of the linear system.

4. Solve the travel problems and cost problems using linear equations.

Previously, we learnt how to use substitution and elimination to solve linear systems.

Another way to find the solution of a linear system is by graphing. If the lines intersect in a single point, then the coordinates of the point are the solution of the linear system.

Math Explorations 1

Use the graph to solve the system and then check your solution algebraically.

$$\begin{cases} x + 2y = 7 \\ 3x - 2y = 5 \end{cases}$$

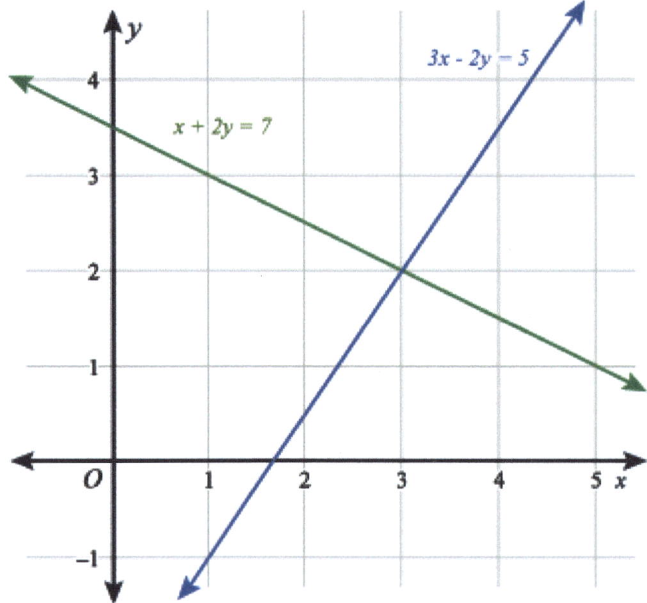

Use the graph to solve the linear system $\begin{cases} -x + y = -7 \\ x + 4y = -8 \end{cases}$. Which of the followings are the solution of the linear system?

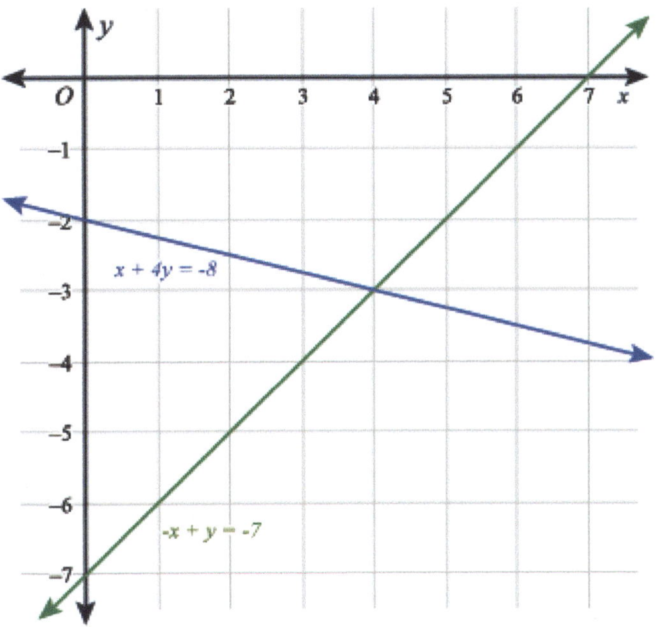

A. $(4, 3)$　　　　B. $(-4, -3)$　　　　C. $(-3, 4)$　　　　D. $(4, -3)$

A linear system can have no solution or infinitely many solutions.

A linear system has no solution when the graphs of the equations are parallel. A linear system with no solution is called an inconsistent system.

For example, the linear system $\begin{cases} 3x + 2y = 10 \\ 3x + 2y = 2 \end{cases}$.

The graphs of the equations are shown below.

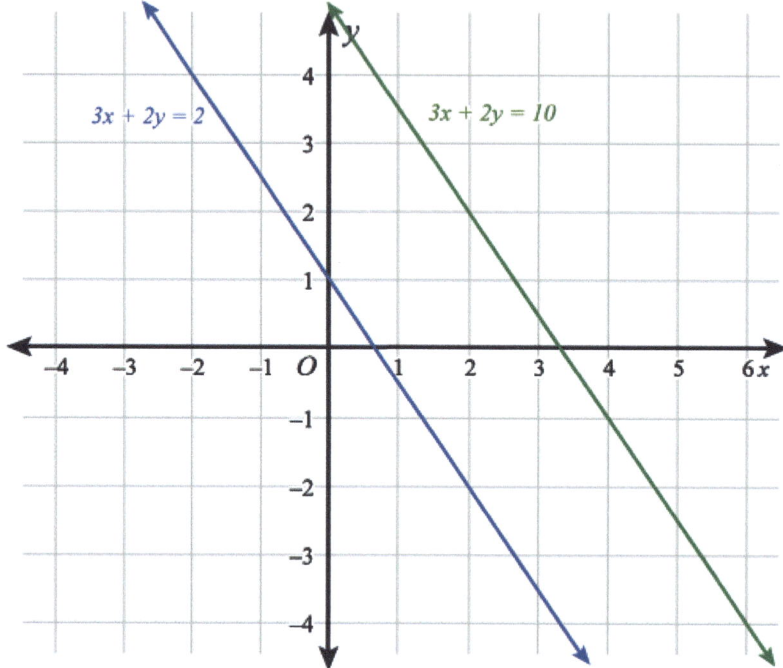

We can find that there is no intersection because the two lines are parallel. Thus, there is no solution.

A linear system has infinitely many solutions when the graphs of the equations are the same line. A linear system with infinitely many solutions is called a consistent dependent system.

For example, the linear system $\begin{cases} x - 2y = -4 \\ y = \frac{1}{2}x + 2 \end{cases}$.

The graphs of the equations are shown below.

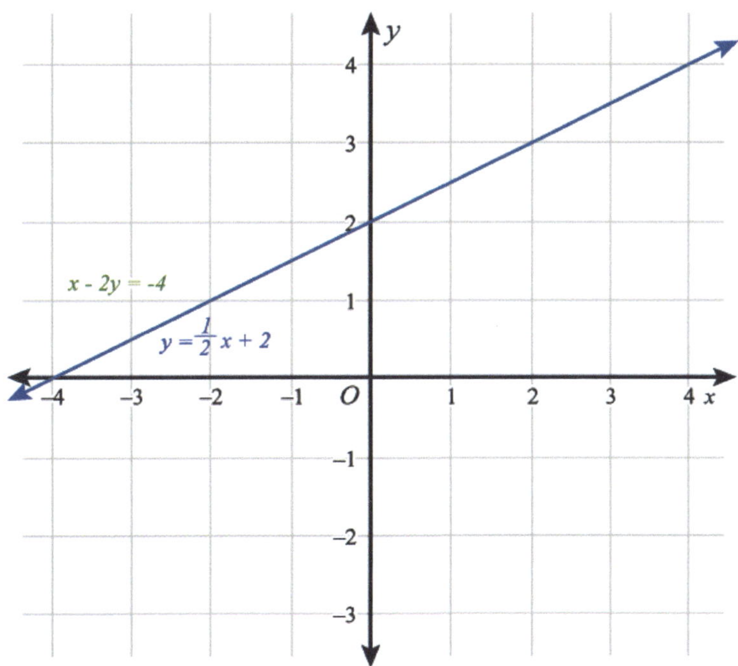

We can find that the two lines are the same line and there are infinitely many intersection. Thus, there are infinitely many solutions.

Math Explorations 2

Identify the number of solutions of the following linear systems.

(1) $\begin{cases} 5x + 3y = 6 \\ -5x - 3y = 3 \end{cases}$

(2) $\begin{cases} y = 2x - 4 \\ -6x + 3y = -12 \end{cases}$

Identify the number of solutions of the following linear systems.

(1) $\begin{cases} 5x + y = -2 \\ -10x - 2y = 4 \end{cases}$

(2) $\begin{cases} 6x + 2y = 3 \\ 6x + 2y = -5 \end{cases}$

Travel problems involve calculating the time, speed (rate), or distance related to the movement of objects. These problems often deal with objects that are moving at constant speeds, allowing us to describe their motion using linear equations with x representing time and y representing distance.

In travel work problems, the slope of a linear equation represents the rate or speed of travel.

The y-intercept in travel work problems may represent an initial distance or starting point, depending on the context.

When two or more objects are moving in different directions, at varying speeds, and/or with distinct initial distances, the point of intersection of the linear equations that describe their motion represents the time and distance at which the objects meet each other.

Math Explorations 3

As shown in the figure, l_1 and l_2 are the graphs of linear equations which represent the trips of John and Sam riding a bicycle, respectively. s and t are the distance and the time they travelled, respectively. The difference between their speeds is ________ km/h.

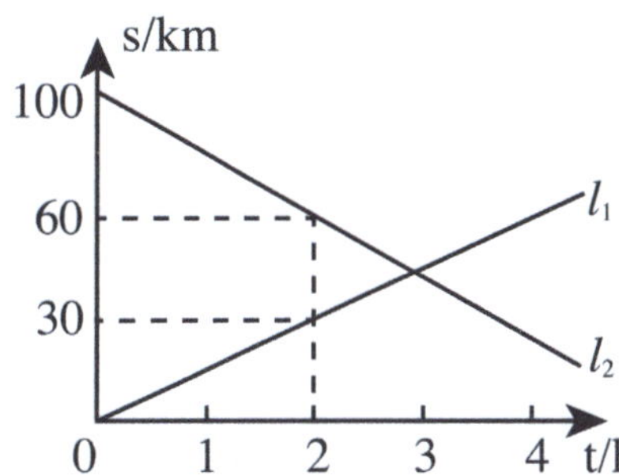

One day, Peter planned to go to the library to read some books and then go back home. The relationship between the distance from home and the time after he left home is shown in the figure below. If Peter spent 30 min to read books in the library, the distance from home after he left home 50 min is _______ km.

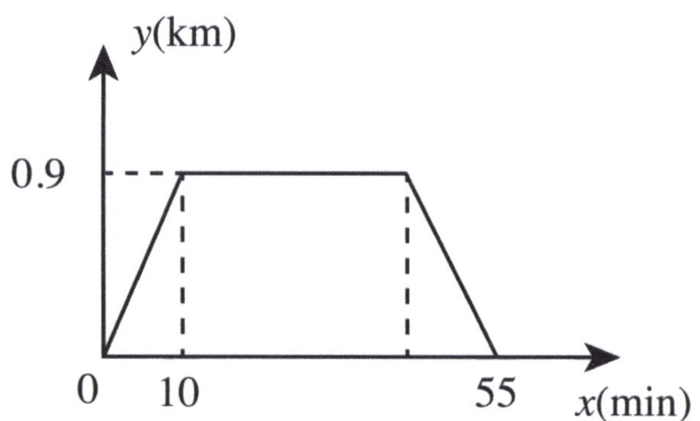

Cost problems can also be modeled using piecewise linear equations, especially when there are different cost structures or rates applicable over different intervals or ranges. Piecewise linear equations allow us to capture these variations and provide a more accurate representation of cost relationships.

In piecewise linear equations that model a "tiered pricing" cost problem, the slopes of the equations represent the price per unit for each tier, while the intercepts may indicate fixed costs or starting points for each interval. A common approach to determining the linear equation for a "pricing tier interval" is to use the point-slope form $y - y_0 = m(x - x_0)$, where m represents the unit cost over that interval, x_0 represents the start of the "pricing tier interval", and y_0 represents the accumulated cost at x_0; obviously, (x_0, y_0) is the end point of the interval.

Math Explorations 4

In order to encourage people to save water, the water use below 3000 tons will be charged for 0.5 cents/ton, and the excess part will be charged for 0.8 cents/ton.
(1) If the use of water is 3200 tons, the water bill is _______ cents; if the use of water is 2800 tons, the water bill is _______ cents.

(2) What is the equation of the relationship between the water bill y and the use of water x?

(3) If the water bill is 1540 cents, the use of water is _______ tons.

1 The standards of the phone bills in city A are:
(1) The phone bill is 3 cents within 3 minutes (including 3 minutes).
(2) When the call time exceeds 3 minutes, the excess part will be charged for 1.1 cents per minute.
In a call, if the call time exceeds 3 minutes, then the relationship between the phone bill y and the call time x is ______ .

2 Amy wants to buy some apples. The figure below shows the relationship between the cost and the amount of purchase. Thus, she buys 3 kg of apples in one time can save ______ cents compared to buy 1 kg of apples three times.

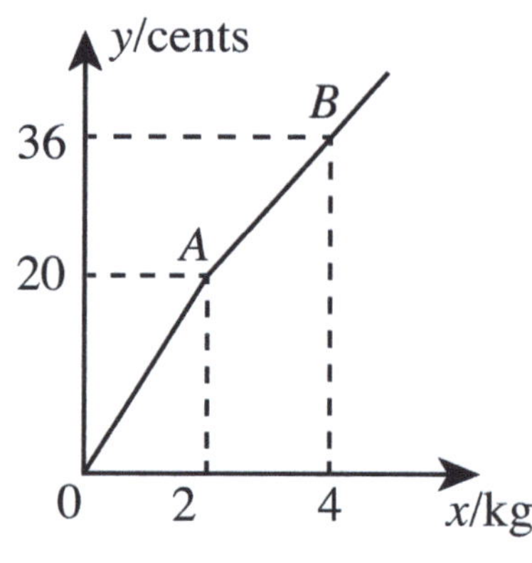

A. 2   B. 4   C. 5   D. 6

1 Which of the following linear systems has no solution?

A. $\begin{cases} 2x - 2y = 6 \\ y - x = -3 \end{cases}$
B. $\begin{cases} 3x + 4y = 1 \\ 2x - 5y = 16 \end{cases}$
C. $\begin{cases} 2x + y = 2 \\ 4x + 2y = -4 \end{cases}$
D. $\begin{cases} 2x + y = 2 \\ x + y = 3 \end{cases}$

2 Villages A, B and C is located beside a straight road. Tom and Jim travel toward village C from village A and village B respectively. The relationship between the distance from village C and time is shown in the figure. Answer the following questions.

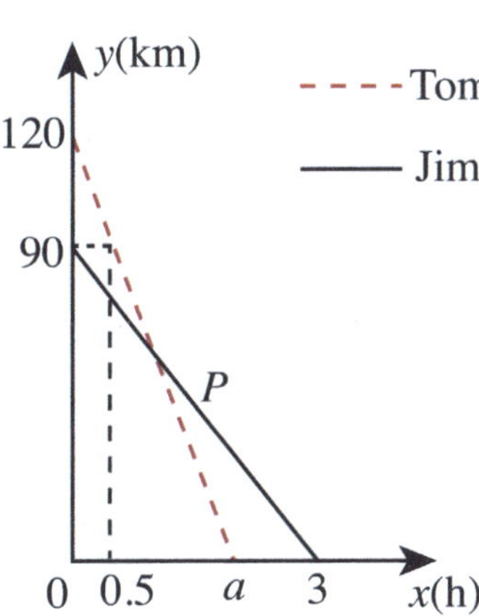

(1) The distance from village A and village C is _______ km, a = _______ .

(2) They meet each other after _______ hour(s).

Summary:

Mind Map

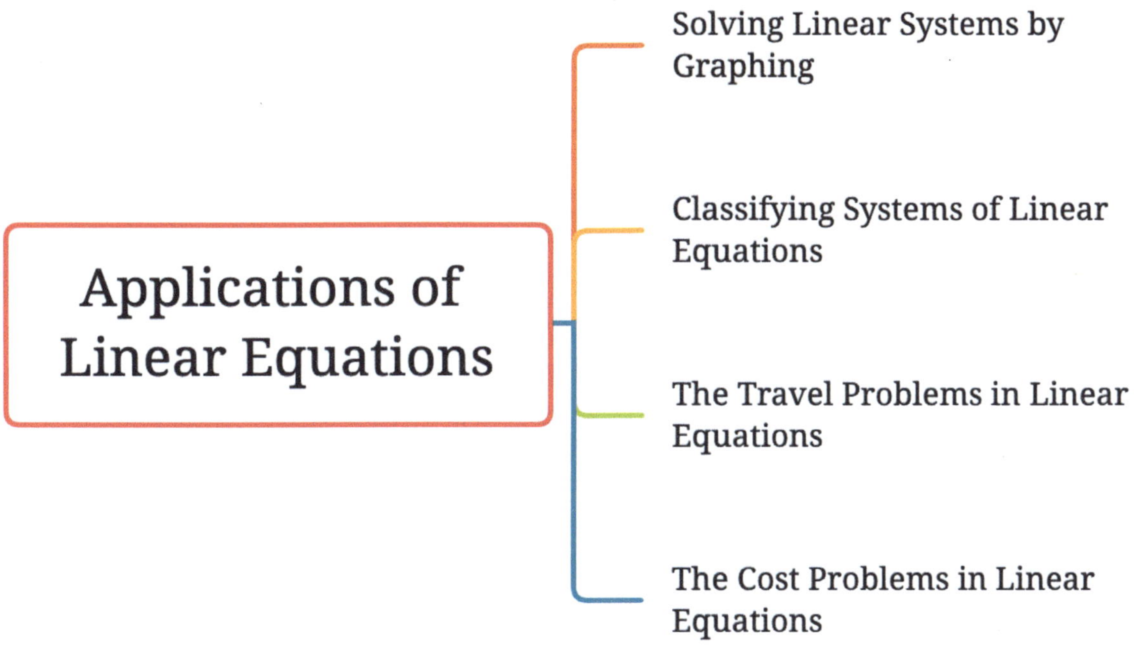
Applications of Linear Equations
Solving Linear Systems by Graphing
Classifying Systems of Linear Equations
The Travel Problems in Linear Equations
The Cost Problems in Linear Equations

1 Use the graph to solve the system and then check your solution algebraically.

$$\begin{cases} 2x - y = 3 \\ 3x + 2y = 1 \end{cases}, (\underline{\hspace{2em}}, \underline{\hspace{2em}}).$$

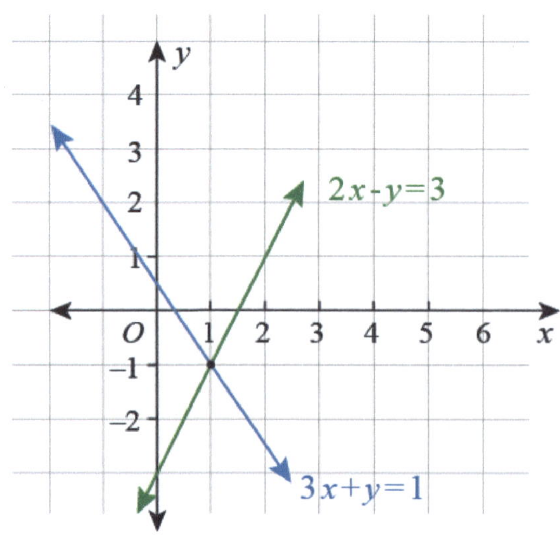

2 Use the graph to solve the linear system $\begin{cases} x + 3y = 5 \\ 3x - 5y = 1 \end{cases}$. Which of the followings are the solution of the linear system?

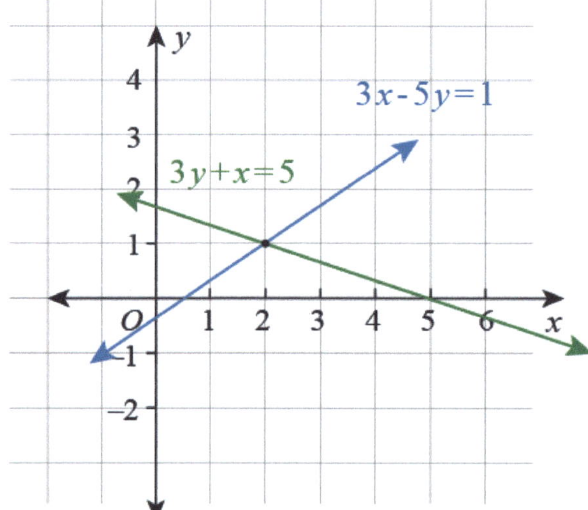

A. $(1, 2)$ B. $(-1, -2)$ C. $(-2, 1)$ D. $(2, 1)$

2024 Fall Algebra 1B A Lesson 1-8

3 Which of the following linear systems has no solution?

A. $\begin{cases} 3x - y = 6 \\ y - x = 2 \end{cases}$
B. $\begin{cases} 4x + 3y = 1 \\ 6x - 3y = 2 \end{cases}$
C. $\begin{cases} 2x + 3y = 2 \\ 4x + 6y = -4 \end{cases}$
D. $\begin{cases} 2x - 2y = 2 \\ x + y = 1 \end{cases}$

4 Identify the number of solutions of the following linear systems.

(1) $\begin{cases} 3x + y = -4 \\ -9x - 3y = 12 \end{cases}$
(2) $\begin{cases} 2x + 5y = 1 \\ 2x + 5y = -1 \end{cases}$

5 Identify the number of solutions of the following linear systems.

(1) $\begin{cases} x + 2y = 4 \\ -x - 2y = -4 \end{cases}$
(2) $\begin{cases} y = 3x - 5 \\ -6x + 2y = -18 \end{cases}$

6 James plans to take a walk. He decides to go to the stadium first, then go to the supermarket, and then go back home. The speed of James is constant. The figure below shows the relationship between the distance from home and the time after he leaves home. Answer the following questions.

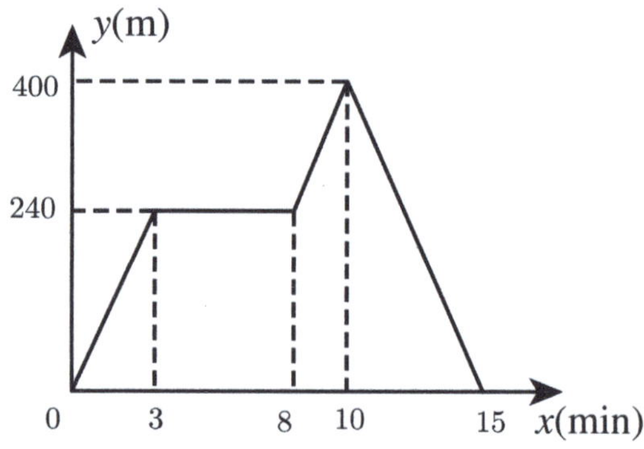

(1) The speed of James is ______ m/min.

(2) After James leaves the supermarket on his way home, the equation of the relationship between y and x is ______ .

7 Tom and Sam participated in a 400-meter running test and they finished at the same time. The figure below shows the relationship between the distance and the time. After ______ seconds, they first meet.

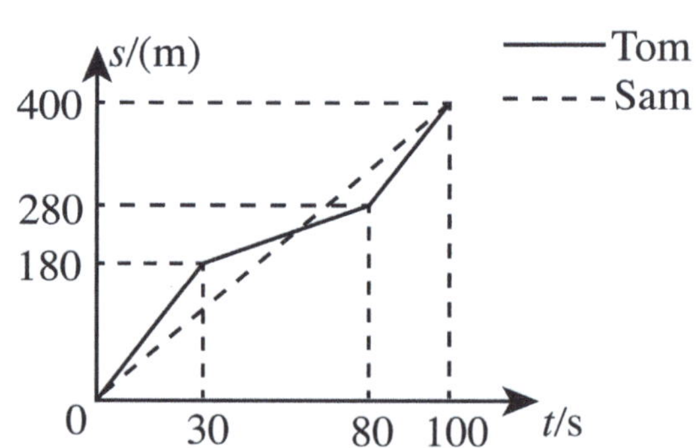

8 The figure below shows the relationship between the total cost of paper and the amount of paper (x). From the figure, if Peter buys 120 pieces of paper, the total cost of paper is _______ cents.

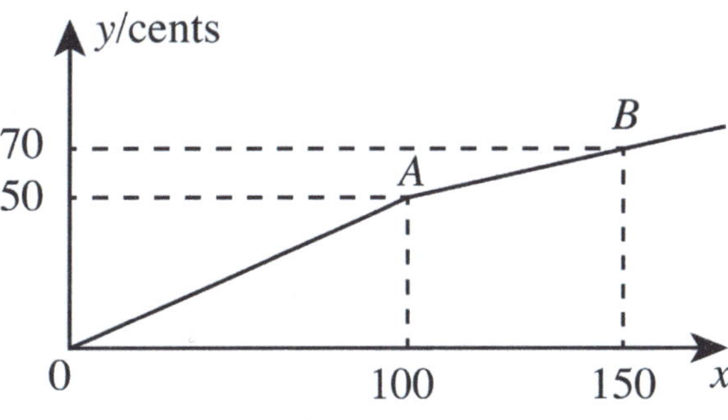

9 The standards of the taxi fare in city A are:
(1) The taxi fare is 5 dollars within 3 miles (including 3 miles).
(2) When the mileage exceeds 3 miles, the excess part will be charged for 1.2 dollars/mi.
Assume that the taxi fare is y (dollars) and the mileage is x (miles). When $x > 3$, the equation of the relationship between x and y is _______ .

10 The standards of the electricity rate in city A are:

A. When the use of electricity does not exceed 180 kWh, the electricity rate will be charged for 1.5 cents/kWh.

B. When the use of electricity exceeds 180 kWh but does not exceed 280 kWh, the excess part will be charged for 1.8 cents/kWh.

C. When the use of electricity exceeds 280 kWh, the excess part will be charged for 2 cents/kWh.

(1) If the use of electricity is 160 kWh, the electricity cost is _______ cents, and if the use is 200 kWh, the electricity cost is ______ cents.

(2) Assume that the use of electricity is x (kWh), the electricity cost is y (dollars). The equation of the relationship between x and y is ______ .

Lesson 7
Graphing Linear Inequalities in Two Variables

1. Use graphs to represent the solution of the linear inequality in two variables.

2. Draw the solution of the system of linear inequalities in the coordinate plane.

3. Find the area of the triangle enclosed by the graphs of the system of linear inequalities.

1. Identify the corresponding equation by changing the inequality to an equality. For example, the corresponding linear equation of $ay \geq bx + c$ is $ay = bx + c$ (a, b, c are constants).

2. Graph the corresponding linear equation by finding two points on the line.

3. Determine the shading region. Identify whether the inequality is greater than ($>, \geq$) or less than ($<, \leq$), and whether it is inclusive ($\geq, \leq$) or exclusive ($>, <$). This will determine whether the shading should be above or below the boundary line and whether the boundary line should be solid or dashed.

Inequality	Shading	Boundary Line
$>$	Above	Dashed
$<$	Below	Dashed
$\geq$	Above	Solid
$\leq$	Below	Solid

4. Check a test point: To confirm that the shading is correct, select a point within the shaded region and substitute its coordinates into the original inequality. If the inequality holds true for the chosen point, then the shading is correct. If not, double-check the graph and shading.

Math Exploration 1

1. Which of the following ordered pairs is a solution of $x + 2y > 3$?
   A. $(0, -2)$　　　　B. $(5, -4)$　　　　C. $(1, 1)$　　　　D. $(2, 1)$

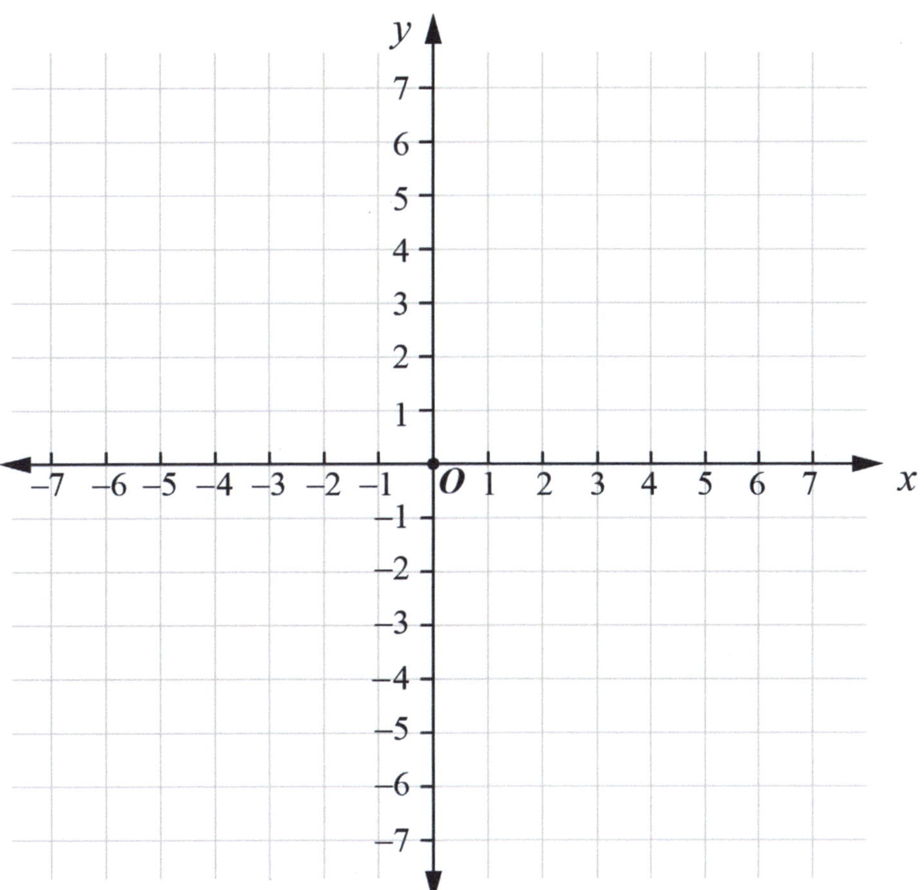

3 Graph the inequality $y > -x + 1$.

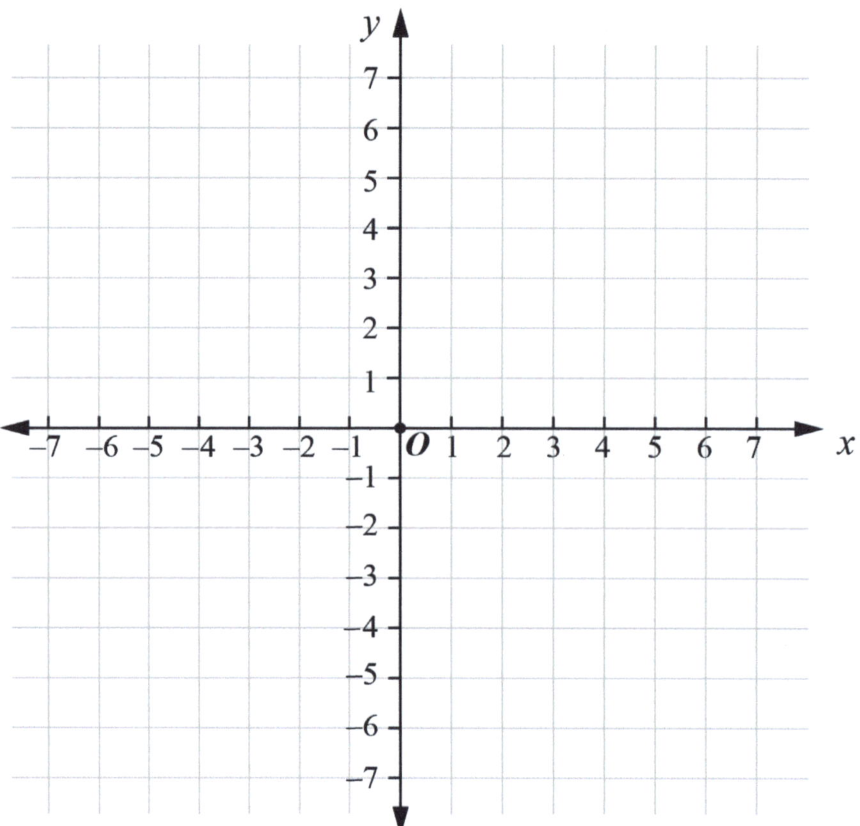

1 Which of the following ordered pairs is a solution of $4x - 2y > 8$?

A. $(0, 0)$ B. $(-5, 4)$ C. $(3, 1)$ D. $(1, 2)$

2 Graph the inequality $x > 2$.

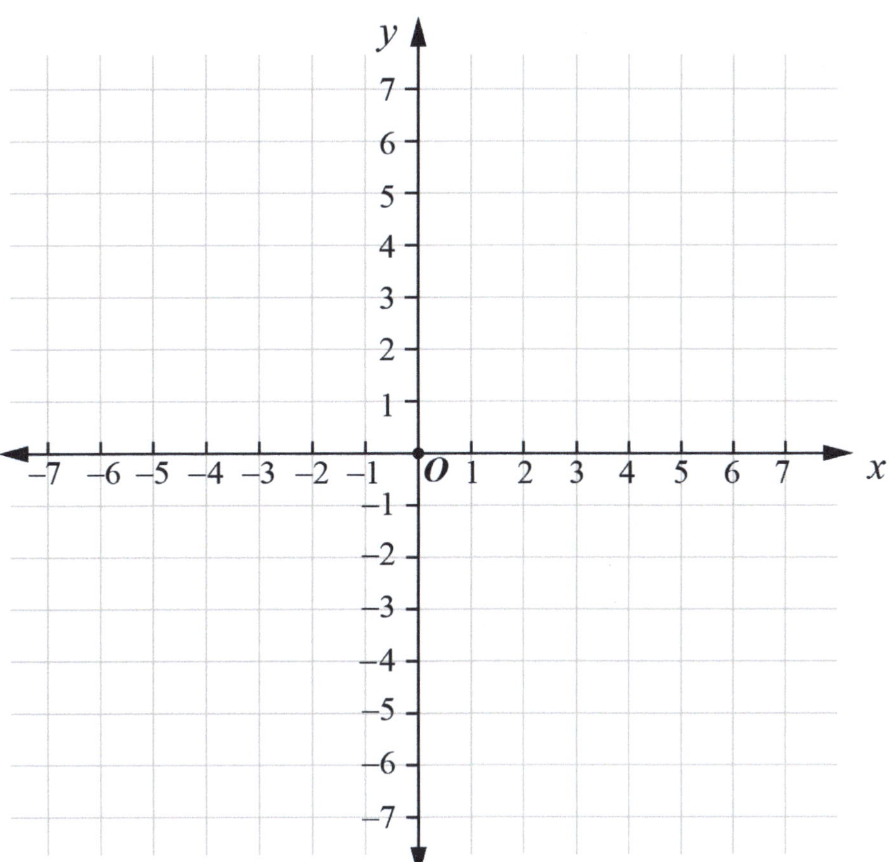

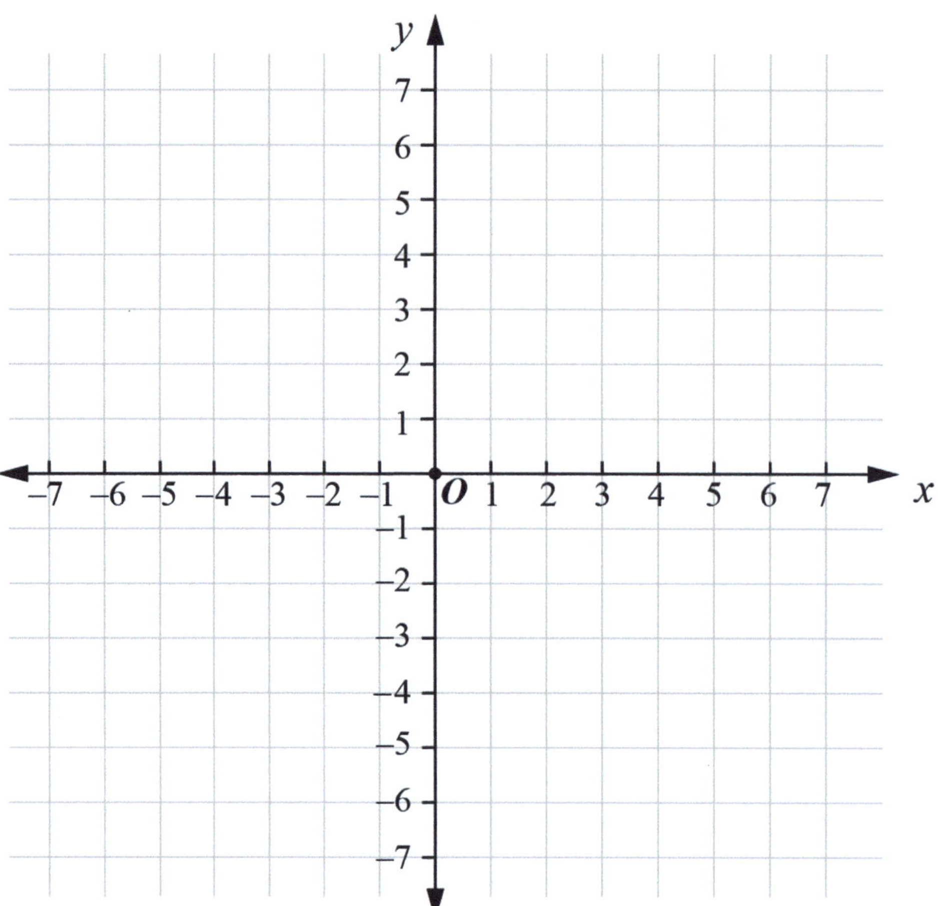

Math Exploration 2

Graph the following compound inequalities.

(1) $\begin{cases} x < -2 \\ y \leqslant 1 \end{cases}$

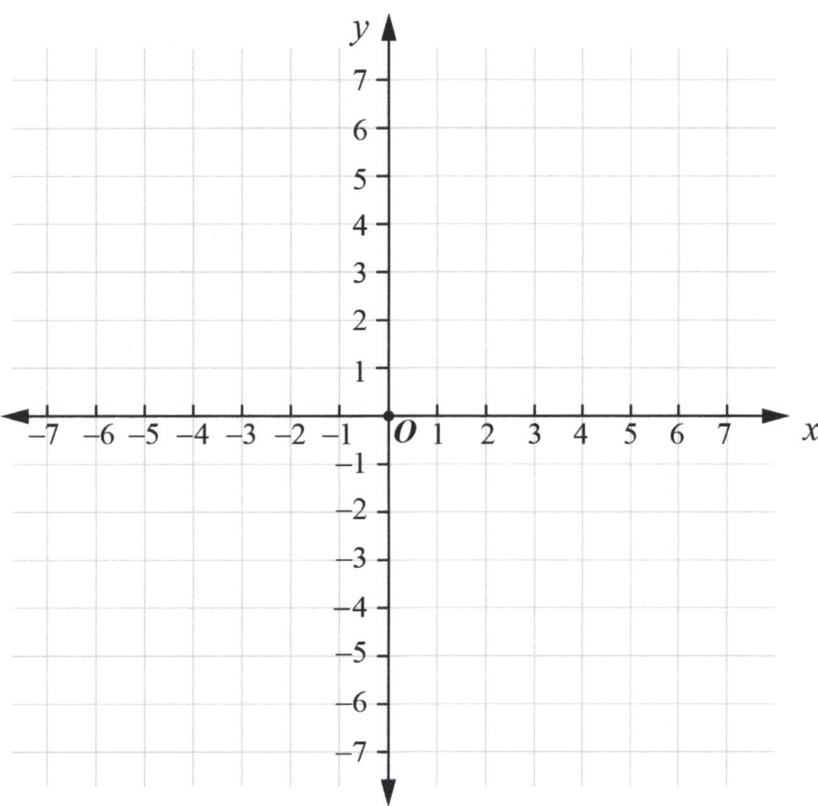

(2) $\begin{cases} x - y + 1 \leqslant 0 \\ 2x - y - 2 \geqslant 0 \end{cases}$

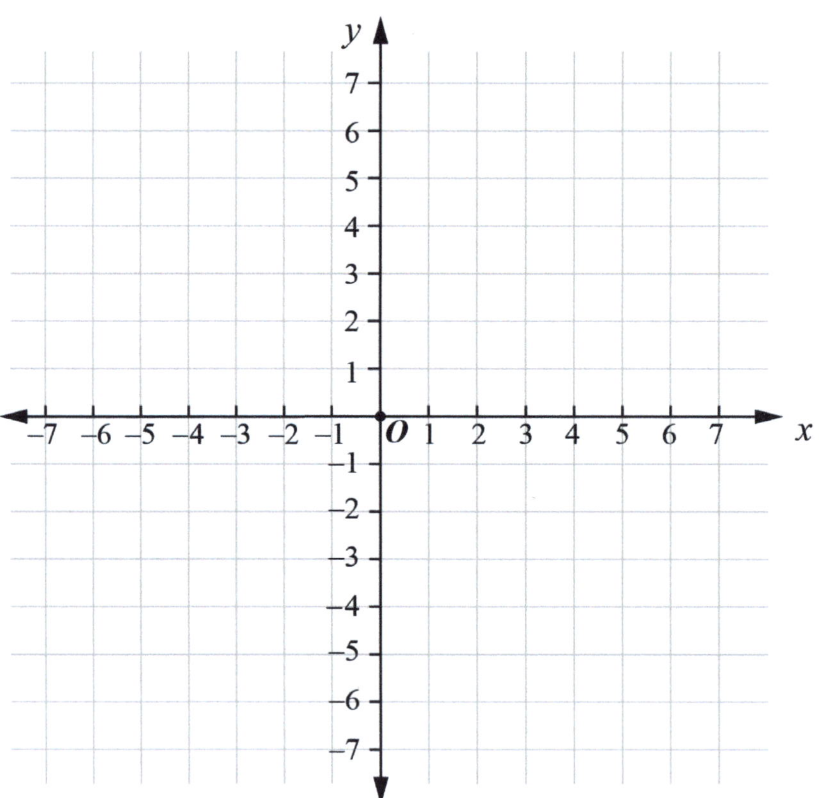

Graph the system of linear inequalities $\begin{cases} x + y \geqslant 1 \\ x - y \leqslant 2 \end{cases}$.

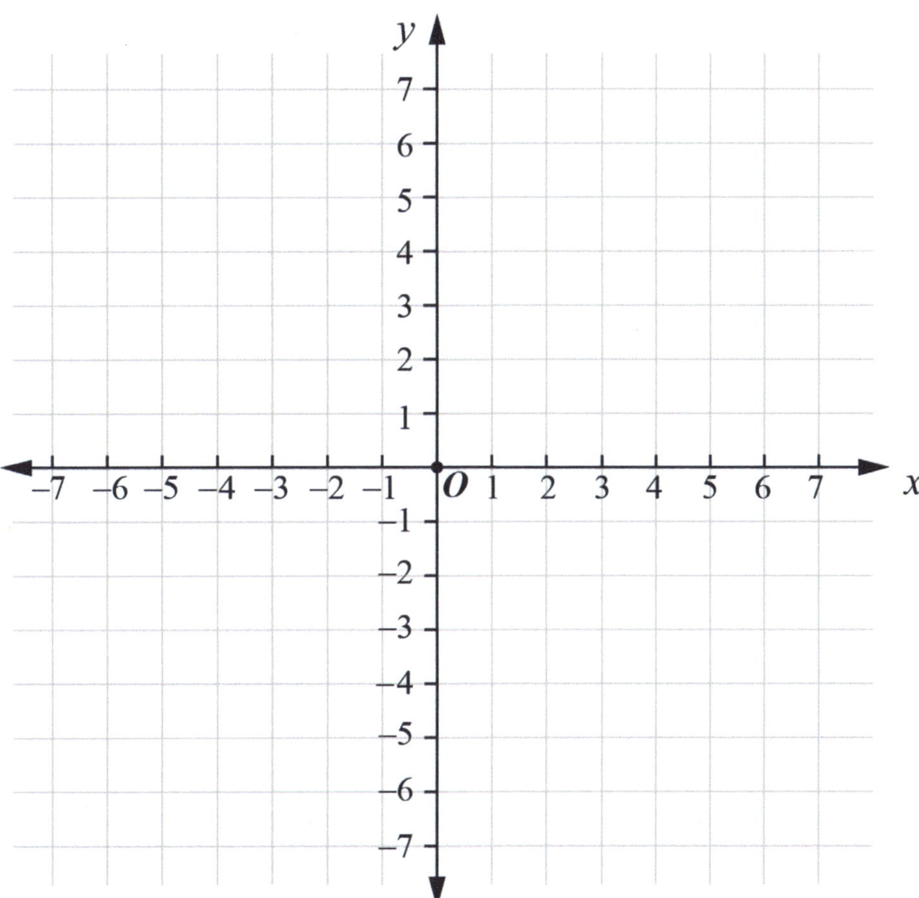

Math Exploration 3

Graph the system of linear inequalities $\begin{cases} x - y < 3 \\ x + y < 3 \\ x > -2 \end{cases}$.

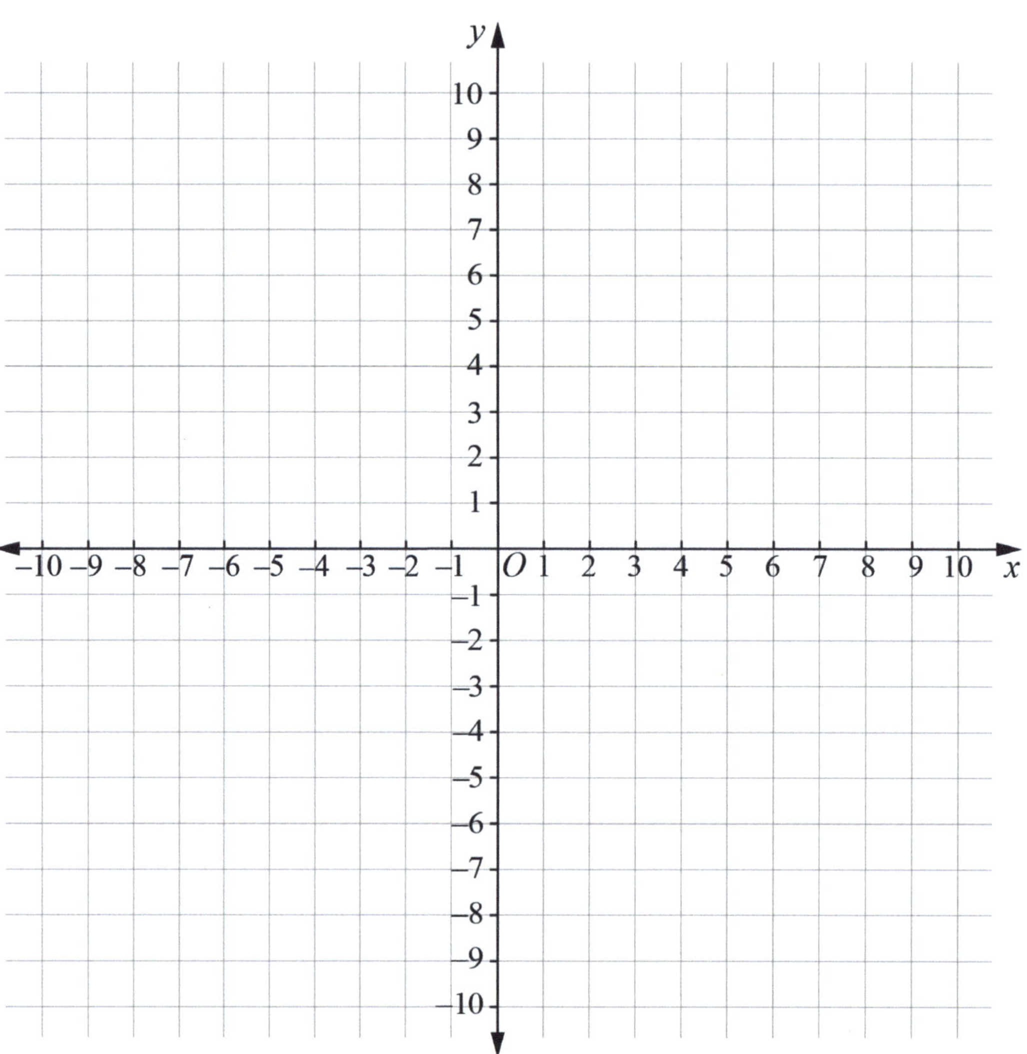

Graph the system of linear inequalities $\begin{cases} x + y \geqslant 2 \\ 2x - y \leqslant 4 \\ x - y > 0 \end{cases}$.

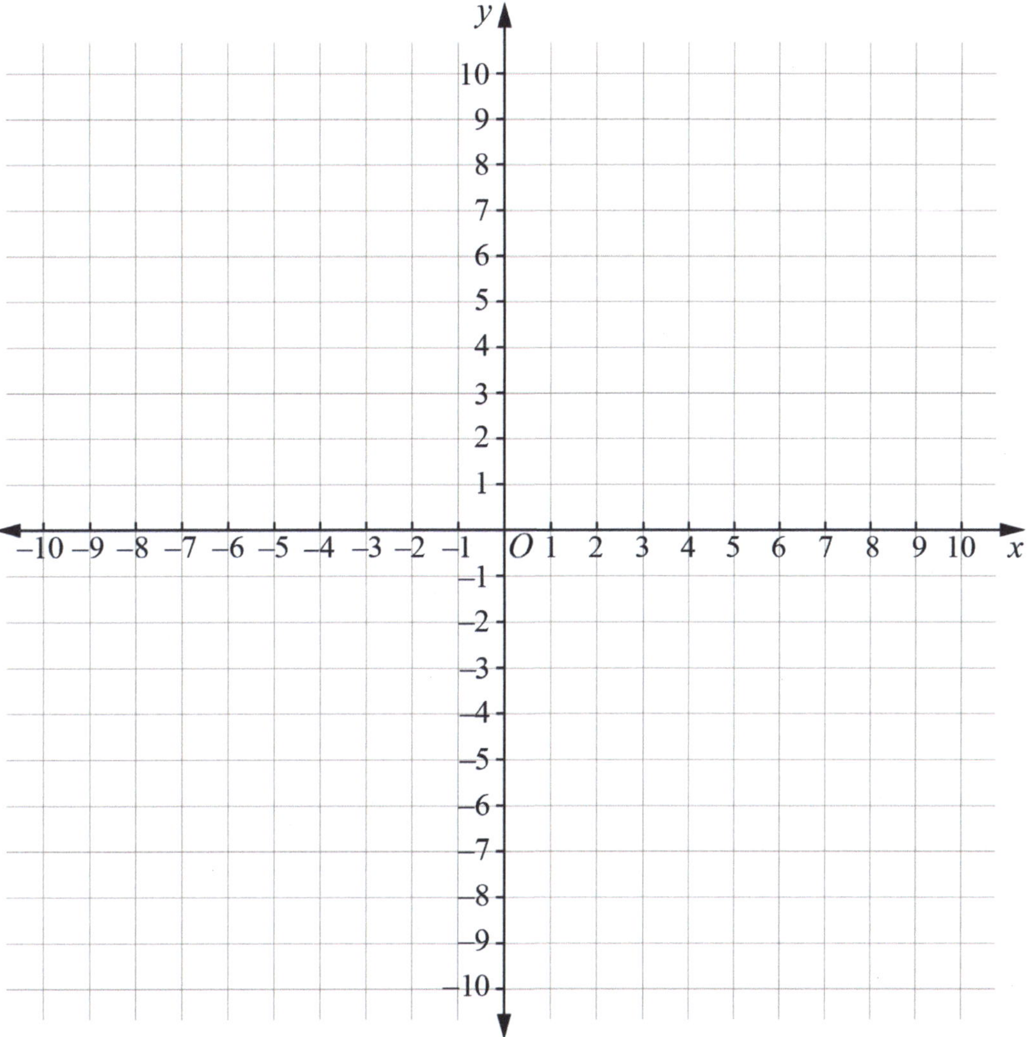

Math Exploration 4

Find the area of the triangle enclosed by the system of linear inequalities
$$\begin{cases} x - y + 6 \geqslant 0 \\ x + y \geqslant 0 \\ x \leqslant 3 \end{cases}.$$

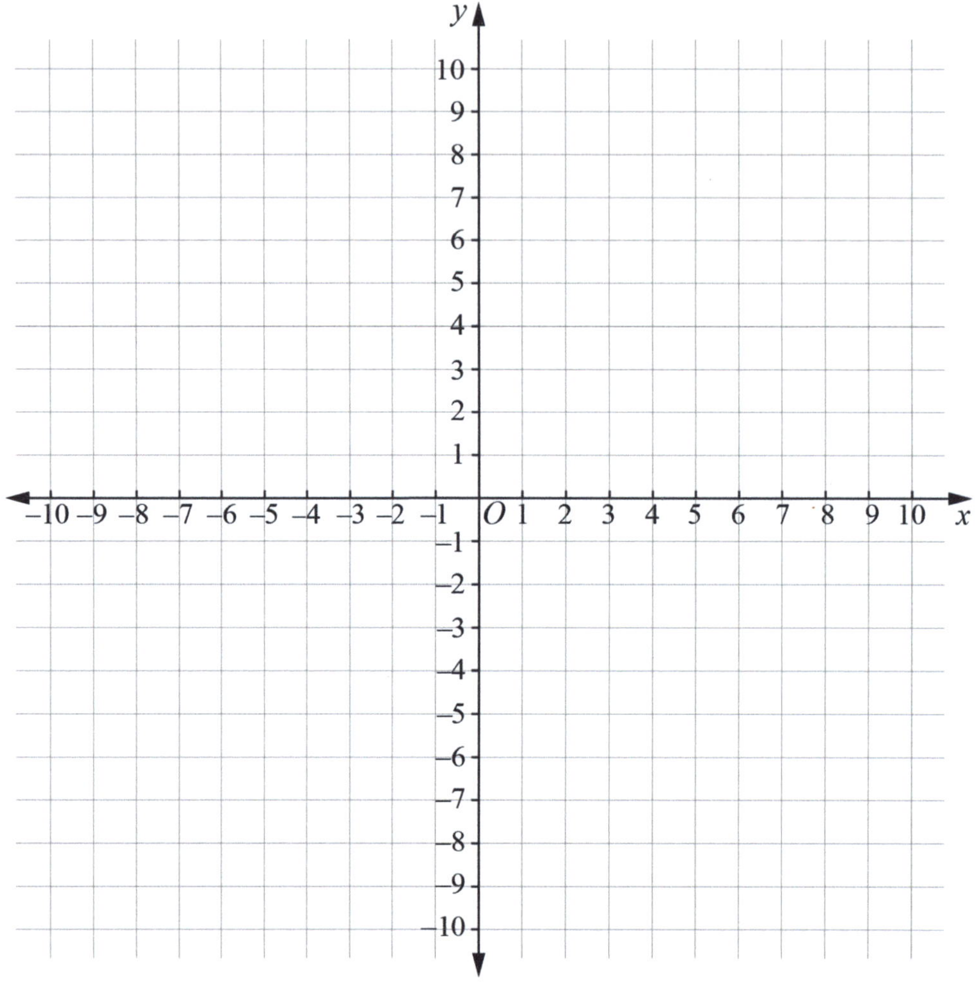

The area of the triangle enclosed by the system of linear inequalities $\begin{cases} 2x + y - 2 \geqslant 0 \\ x - y \leqslant 0 \\ y \leqslant 2 \end{cases}$

is ______ .

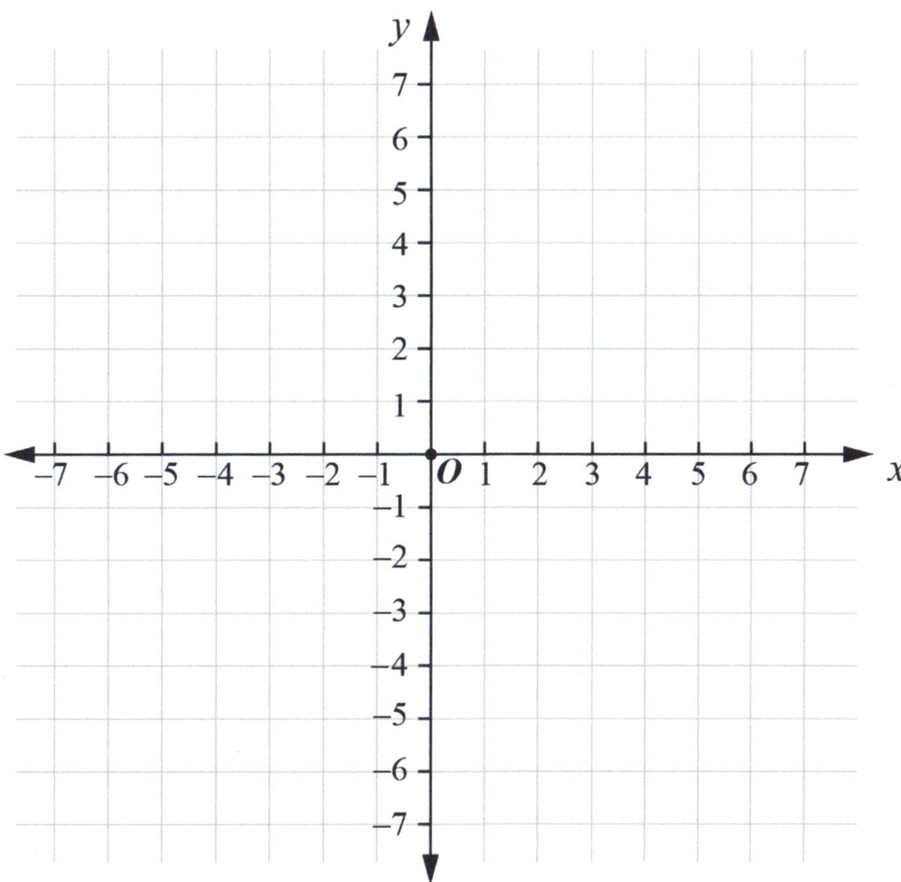

1 Which of the following ordered pairs is a solution of $5x + y > 5$?

A. $(0, 1)$ 　　　　 B. $(0, 5)$ 　　　　 C. $(1, 1)$ 　　　　 D. $(1, 0)$

2 The area of the triangle enclosed by the system of linear inequalities $\begin{cases} x + y \geqslant 2 \\ x \leqslant 1 \\ y \leqslant 2 \end{cases}$

is ______ .

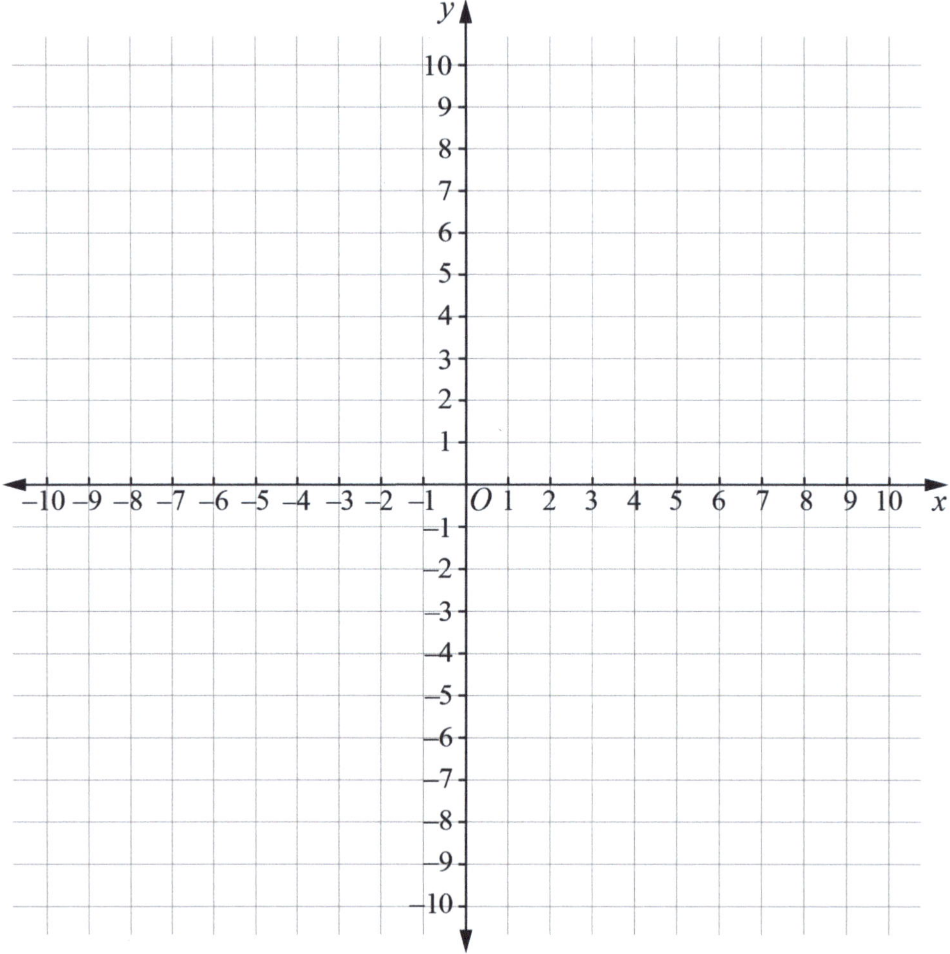

Summary:

Mind Map

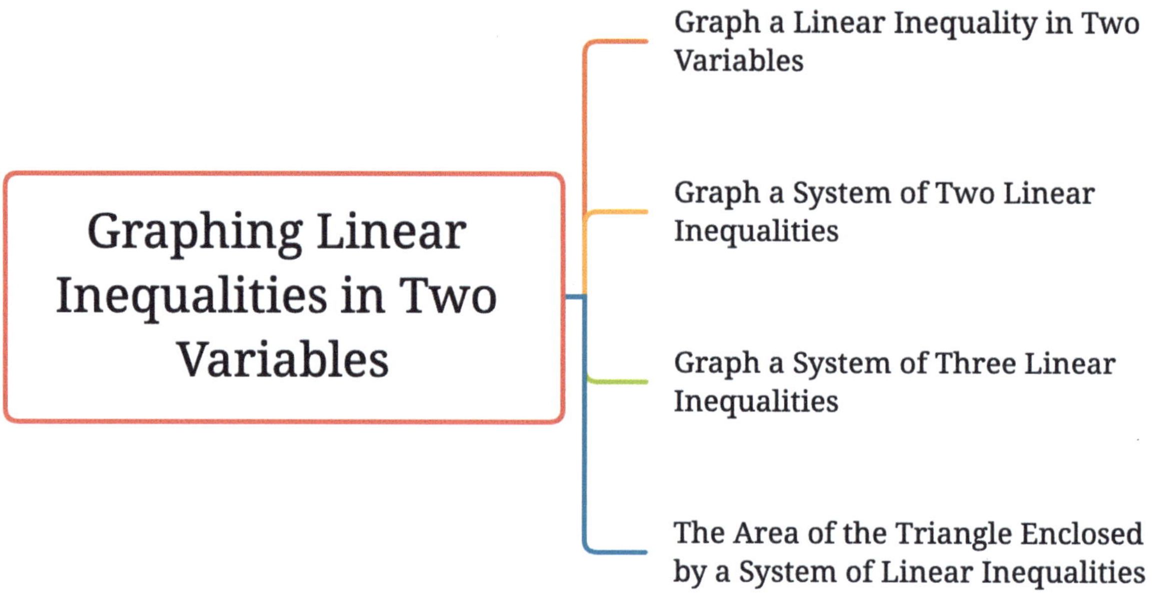
Graphing Linear Inequalities in Two Variables
Graph a Linear Inequality in Two Variables
Graph a System of Two Linear Inequalities
Graph a System of Three Linear Inequalities
The Area of the Triangle Enclosed by a System of Linear Inequalities

1 Solve the following inequalities.

(1) $-3x \geqslant 2x - 15$

(2) $-2(3x + 2) \leqslant 4x - 10$

2 Solve the following inequalities.

(1) $3(2x - 1) + 3 < 2(x + 2) - 8$

(2) $-2 < \dfrac{4 - 5x}{3} \leqslant 3$

3 Which of the following ordered pairs is a solution of $-3x + y > -2$?

    A. $(1, -2)$ B. $(3, 7)$ C. $(-1, -6)$ D. $(1, 3)$

4 Graph the inequality $x + 4y < 4$.

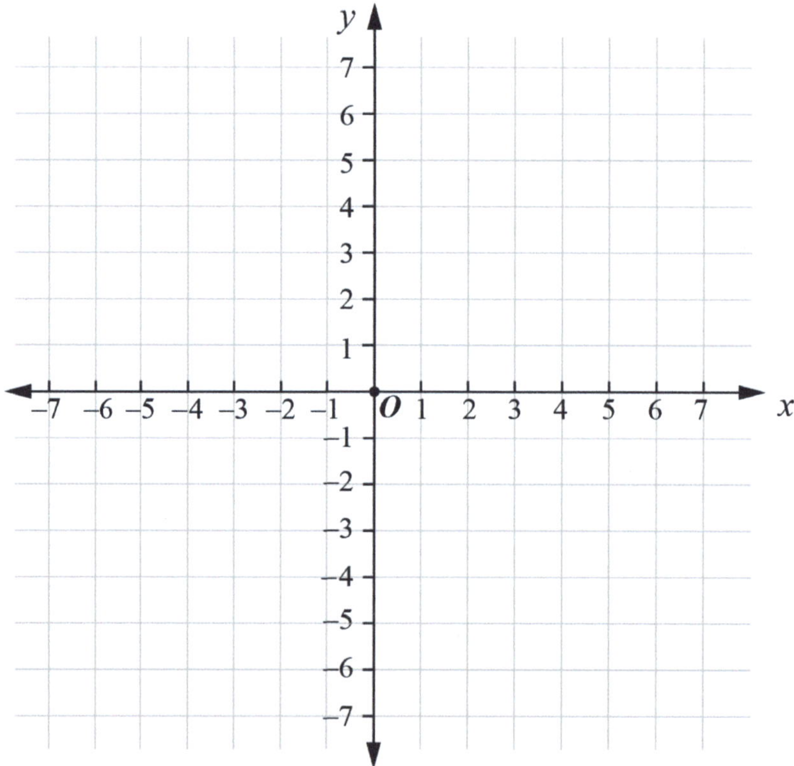

2024 Fall Algebra 1B A Lesson 1-8

5 Graph the following systems of inequalities.

(1) $\begin{cases} x \geq 1 \\ y > 3 \end{cases}$

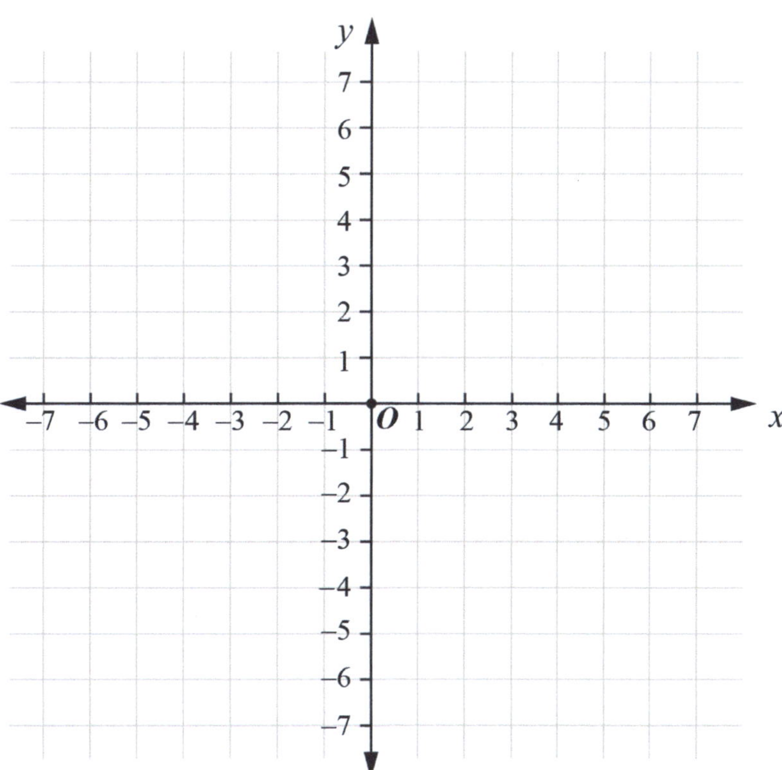

(2) $\begin{cases} x + y + 1 \leqslant 0 \\ x - y + 2 \geqslant 0 \end{cases}$

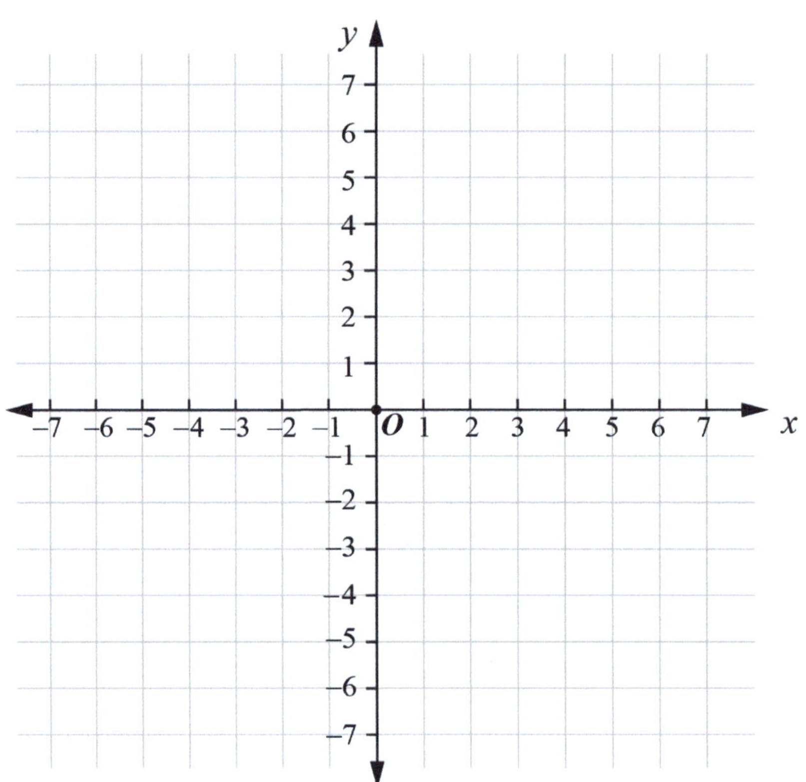

6 Graph the system of linear inequalities $\begin{cases} x + y \geqslant 1 \\ x - 2y \leqslant 4 \end{cases}$.

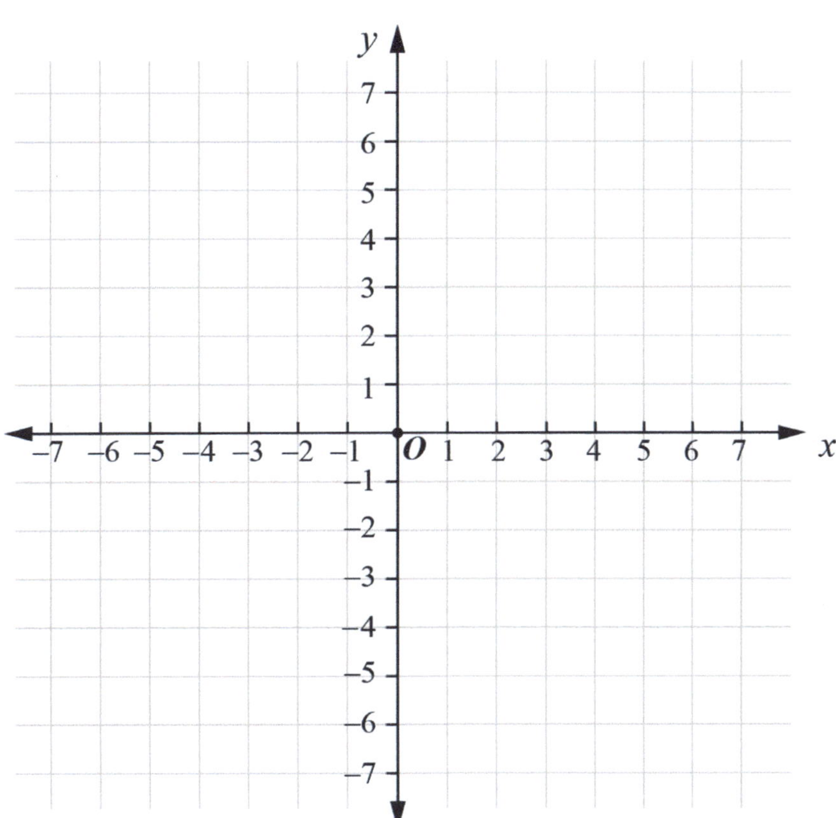

7 Graph the system of linear inequalities $\begin{cases} x - 1 \geqslant 0 \\ y \geqslant 0 \\ x + y - 5 \leqslant 0 \end{cases}$.

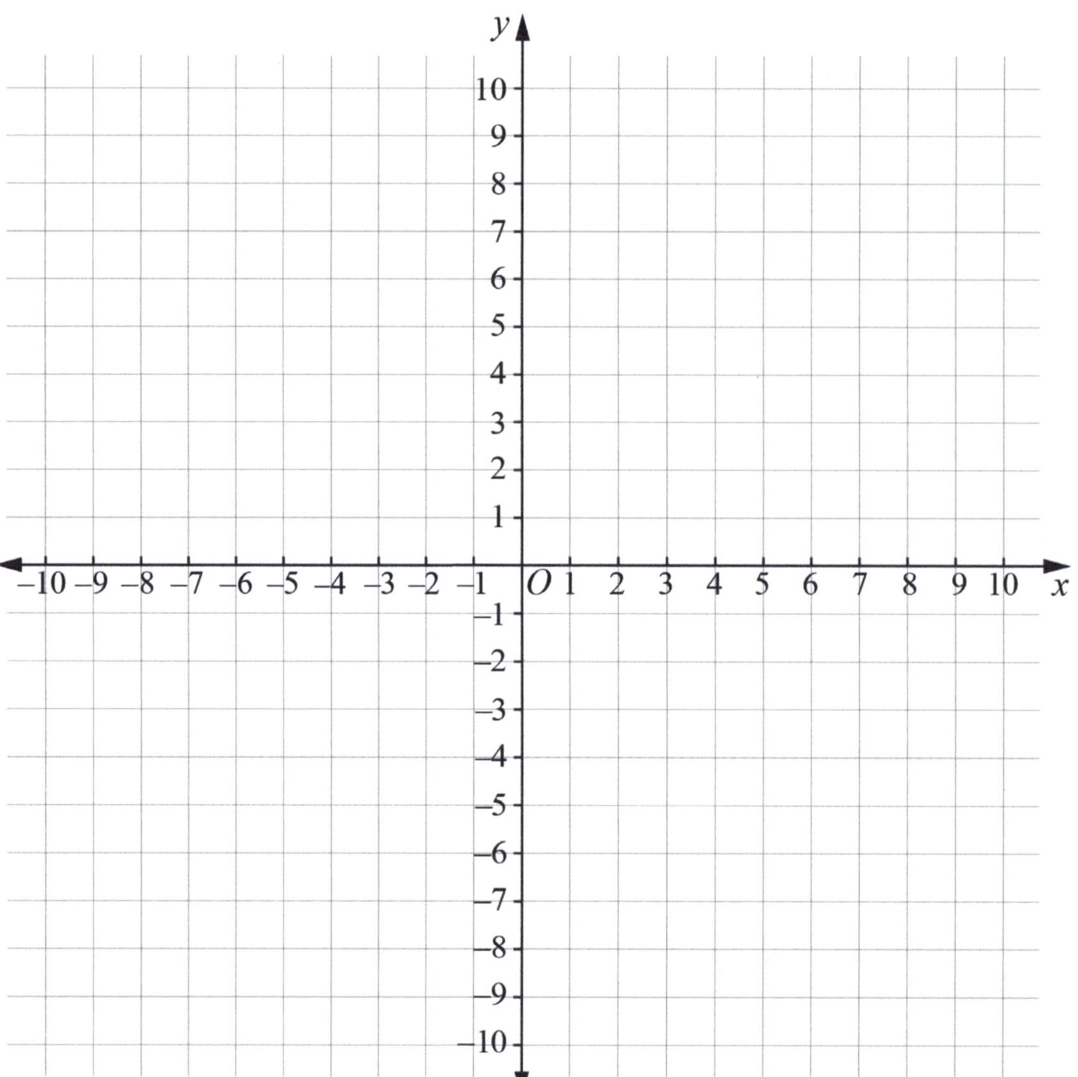

2024 Fall Algebra 1B A Lesson 1-8

8 The area of the triangle enclosed by the system of linear inequalities $\begin{cases} x + 3y - 3 \leqslant 0 \\ x \geqslant 0 \\ y \geqslant 0 \end{cases}$

is ______ .

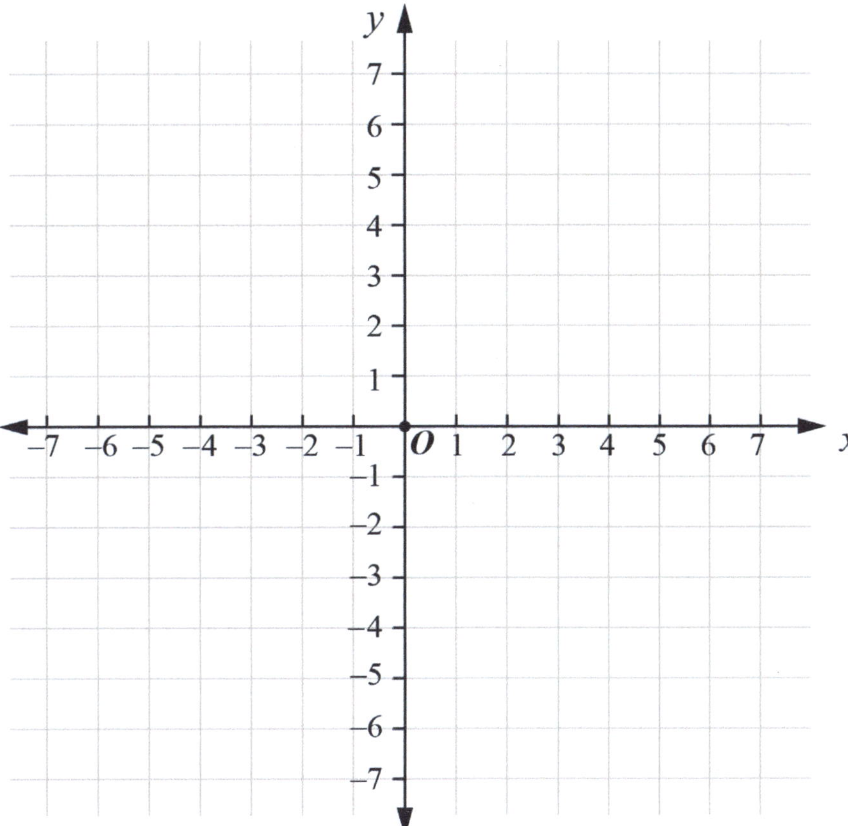

9 The area of the triangle enclosed by the system of linear inequalities $\begin{cases} x - y + 1 \geqslant 0 \\ x - 5 \leqslant 0 \\ y - 3 \geqslant 0 \end{cases}$

is ______ .

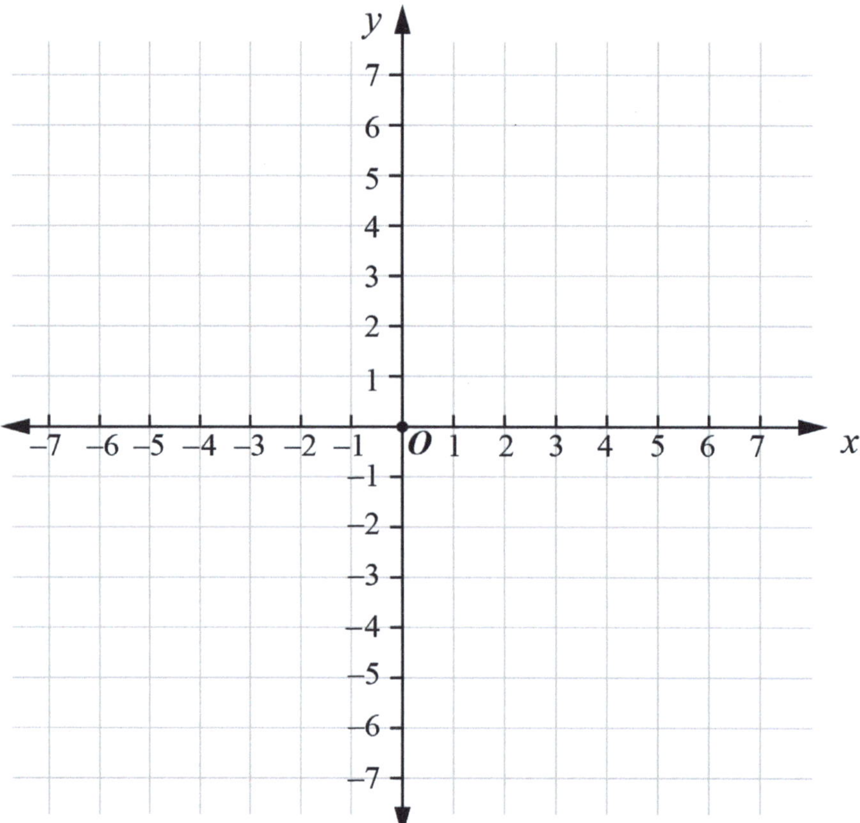

10 The area of the triangle enclosed by the system of linear inequalities $\begin{cases} x + y - 2 \geqslant 0 \\ x - y + 1 \leqslant 0 \\ y - 3 \leqslant 0 \end{cases}$

is ______ .

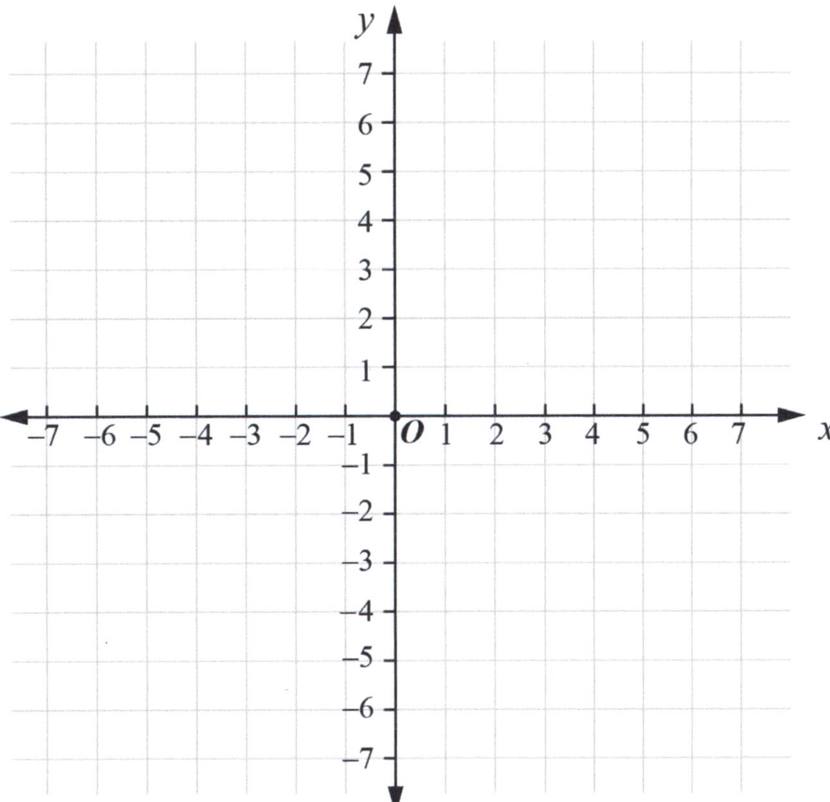

Lesson 8
Mid-Term Review

1. Review on how to solve absolute-value equations and inequalities.

2. Review on how to solve systems of linear equations and inequalities.

3. Review on the basics of linear equations including equation forms and graphs.

4. Review on the applications of linear equations.

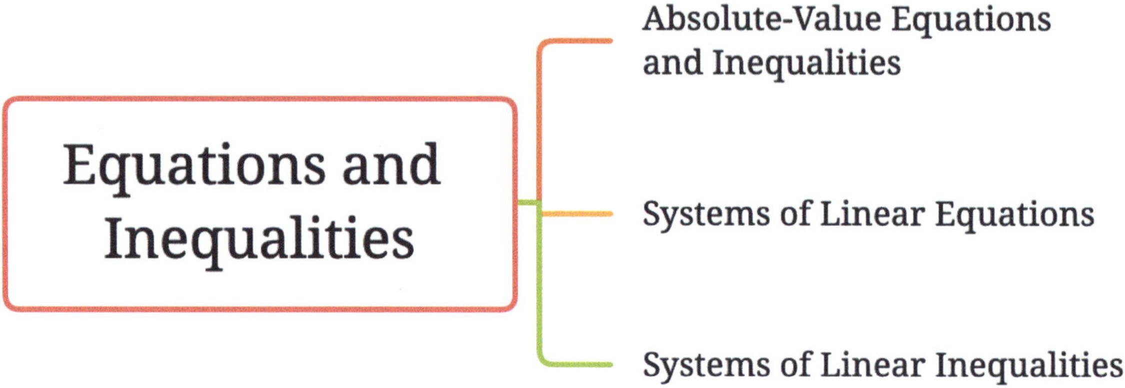

Equations and Inequalities
Absolute-Value Equations and Inequalities
Systems of Linear Equations
Systems of Linear Inequalities

Math Exploration 1

1 Solve the absolute-value equation: $2|3x + 1| = 8$.

2 Solve the following absolute-value inequalities.

(1) $|3x - 3| > 5$

(2) $|2x + 2| \leqslant 6$

Solve the following absolute-value inequality and absolute-value equation.

(1) $5|2x + 13| = 15$

$x = $ ______ or $x = $ ______ .

(2) $|3x + 5| > 8$

Key Points

There are two methods to solve systems of linear equations.

Substitution:

Express one variable with the other one, and substitute the variable in the equation to eliminate one variable.

Elimination:

Add or Subtract the equations to eliminate one variable.

Math Exploration 2

Solve the system of linear equations: $\begin{cases} x + 2y = 3 \\ x - 2y = 1 \end{cases}$.

Solve the system of linear equations: $\begin{cases} 2x + y = 7 \\ 2x - 3y = 3 \end{cases}$.

$x =$ _______ , $y =$ _______ .

Key Points

There are four situations for the solution of the compound inequalities with "AND".

1. If the solution is $\begin{cases} x > a \\ x > b \end{cases}$ $(a < b)$, the final solution is $x > b$.

2. If the solution is $\begin{cases} x < a \\ x < b \end{cases}$ $(a < b)$, the final solution is $x < a$.

3. If the solution is $\begin{cases} x > a \\ x < b \end{cases}$ $(a < b)$, the final solution is $a < x < b$.

4. If the solution is $\begin{cases} x < a \\ x > b \end{cases}$ $(a < b)$, there is no solution for the compound inequalities.

Math Exploration 3

Solve the system of linear inequalities: $\begin{cases} 2x < 6 \\ \dfrac{2x + 1}{3} > x + 1 \end{cases}$.

The solution of $\begin{cases} 2 - x > 1 \\ 5x > 3(x-2) \end{cases}$ is ______ .

A. $x > 1$　　　　B. $-3 < x < 1$　　　　C. $x > -3$　　　　D. No solution

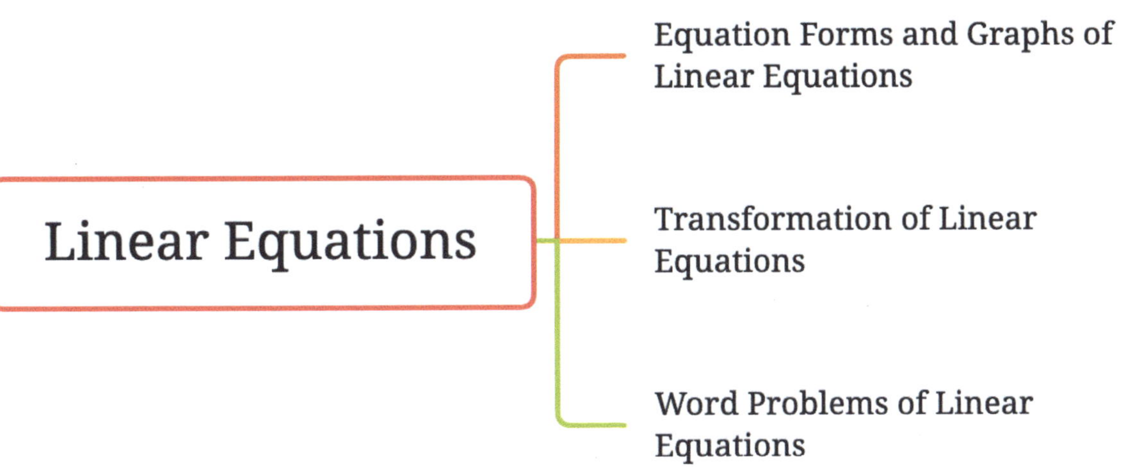
Linear Equations
Equation Forms and Graphs of Linear Equations
Transformation of Linear Equations
Word Problems of Linear Equations

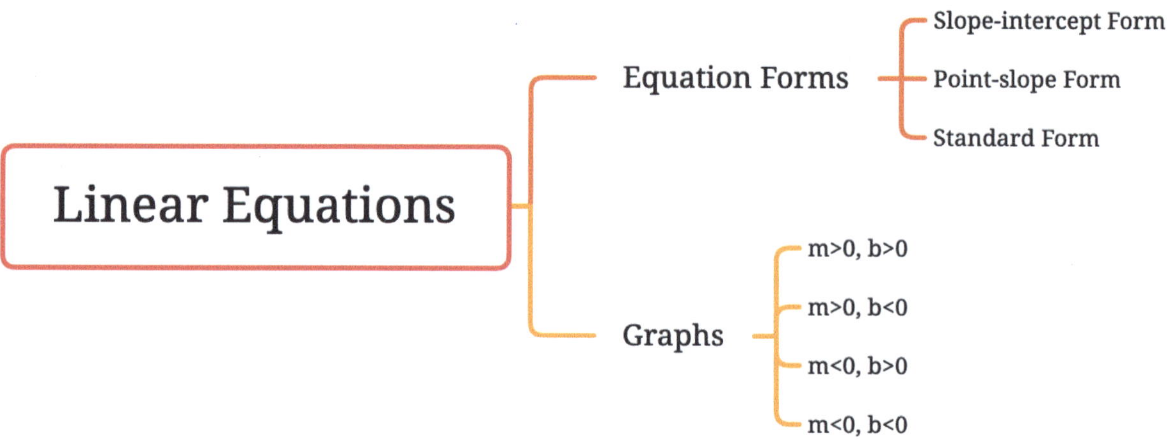

Math Exploration 4

1 There is a line $l_1 : y = mx + b$.

(1) Given that points $(0, 1)$ and $(1, 4)$ are on l_1, write the equation of l_1 in slope-intercept form and convert it to the standard form.

(2) Given that the slope of another line l_2 is 3 and it passes through point $(1, 1)$, write the equation of it in point-slope form and convert it to the slope-intercept form.

(3) What is the distance between the x-intercepts of these two lines? ______ .

2 If the graph of $y = mx + b$ passes through Quadrant I, III and IV, then ______ .
A. $m < 0$ B. $b > 0$ C. $mb < 0$ D. $mb > 0$

Practice 4

1 Given that the graph of the linear equation passes through point $(2, 0)$ and point $(1, -1)$, then the equation of the graph is ______ .
A. $y = x - 2$ B. $y = x + 2$ C. $y = -x - 2$ D. $y = -x + 2$

2 The graph of $y = mx + b$ is shown below, then which of the followings is correct?

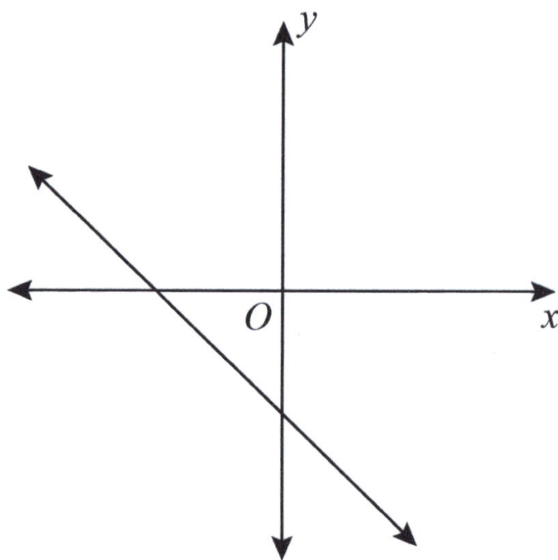

A. $m > 0,\ b < 0$　　　B. $m < 0,\ b < 0$　　　C. $m < 0,\ b > 0$　　　D. $m > 0,\ b > 0$

Key Points

Math Exploration 5

There are two linear equations $l_1 : y_1 = ax + 1$ and $l_2 : y_2 = -3x - b$. Answer the following questions.

(1) If $b = 2$ and the graph of l_2 first moves 3 units to the left and then moves up 2 units, the equation will be $y =$ ______ .

(2) If y_1 and y_2 are symmetric about the x-axis, $a =$ ______ , $b =$ ______ .

(3) If y_1 and y_2 are symmetric about the y-axis, $a =$ ______ , $b =$ ______ .

(4) If y_1 and y_2 are symmetric about the origin, $a =$ ______ , $b =$ ______ .

(5) If y_1 and y_2 are perpendicular, $a =$ ______ .

1 If the graph of the linear equation $y = 3x - 2$ first moves 1 units to the left and then moves down 3 units, the equation will be ______ .
A. $y = 3x - 2$ B. $y = -3x - 2$ C. $y = 3x + 2$ D. $y = -3x + 2$

2 A line that is symmetric to line $y = mx + 6$ about the x-axis passes through $(1, -2)$, then the value of m is ______ .
A. 4   B. −4   C. 8   D. −8

Math Exploration 6

Tim and Ricky travel towards C from A and B repectively. The relationship between the distance from A and time is shown in the figure below. When will they meet each other?

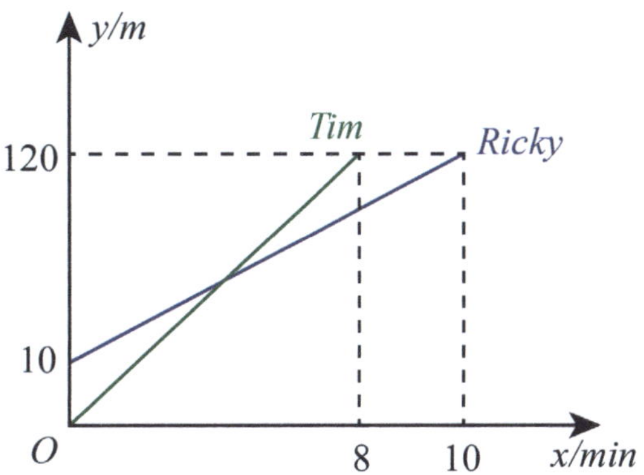

Practice 6

The relationship between the temperature $t(°C)$ and height $h(m)$ is $t = -kh + s$. The observed temperature of 200 m from the ground is 8.4°C, and 500 m from the ground is 6°C. The temperature of 1500 m from the ground is _______ °C.